Crisis Spaces

The financial malaise that has affected the Eurozone countries of Southern Europe – Spain, Portugal, Italy and, in its most extreme case, Greece – has been analysed using mainly macroeconomic and financial explanations.

This book shifts the emphasis from macroeconomics to the relationship between uneven geographical development, financialisation and politics. It deconstructs the myth that debt, both public and private, in Southern Europe is the sole outcome of the spendthrift ways of Greece, Spain, Italy and Portugal, offering a fresh perspective on the material, social and ideological parameters of the economic crisis and the spaces where it unfolded.

Featuring a range of case examples that complement and expand the main discussion, *Crisis Spaces* will appeal to students and scholars of human geography, economics, regional development, political science, cultural studies and social movements studies.

Costis Hadjimichalis is Professor Emeritus in the Department of Geography at Harokopio University of Athens, Greece. He previously held a post in the Department of Urban and Regional Planning at Aristotle University of Thessaloniki, and has been a visiting professor at different universities in Europe, the USA and Australia. His current research and publications concern uneven geographical development, local and regional development, radical geography and landscape analysis. He has been the section editor of the Regional Development section in the *International Encyclopaedia of Human Geography.* Among his recent books are *Space in Left Thought* (co-author Dina Vaiou, 2012 in Greek), *Debt Crisis and Land Dispossession* (2014 in Greek, 2016 in German) and *Geographical Issues Suited to Non-Geographers* (2016 in Greek).

Routledge Studies in Human Geography

This series provides a forum for innovative, vibrant and critical debate within Human Geography. Titles will reflect the wealth of research that is taking place in this diverse and ever-expanding field. Contributions will be drawn from the main sub-disciplines and from innovative areas of work that have no particular sub-disciplinary allegiances.

For a full list of titles in this series, please visit www.routledge.com/series/SE0514

65 Public Urban Space, Gender and Segregation
Women-only urban parks in Iran
Reza Arjmand

66 Island Geographies
Essays and Conversations
Edited by Elaine Stratford

67 Participatory Research in More-than-Human Worlds
Edited by Michelle Bastian, Owain Jones, Niamh Moore and Emma Roe

68 Carceral Mobilities
Interrogating movement in incarceration
Edited by Jennifer Turner and Kimberley Peters

69 Mobilising Design
Edited by Justin Spinney, Suzanne Reimer and Philip Pinch

70 Place, Diversity and Solidarity
Edited by Stijn Oosterlynck, Nick Schuermans and Maarten Loopmans

71 Towards A Political Economy of Resource Dependent Regions
Greg Halseth and Laura Ryser

72 Crisis Spaces
Structures, Struggles and Solidarity in Southern Europe
Costis Hadjimichalis

Crisis Spaces

Structures, Struggles and Solidarity in Southern Europe

Costis Hadjimichalis

LONDON AND NEW YORK

First published in paperback 2020

First published 2018
by Routledge
2 Park Square, Milton Park, Abingdon, Oxon OX14 4RN

and by Routledge
711 Third Avenue, New York, NY 10017

Routledge is an imprint of the Taylor & Francis Group, an informa business

British Library Cataloguing in Publication Data
A catalogue record for this book is available from the British Library

Library of Congress Cataloging-in-Publication Data
Names: Hadjimichalis, Costis, author.
Title: Crisis spaces: structures, struggles and solidarity in Southern Europe / Costis Hadjimichalis.
Description: Abingdon, Oxon; New York, NY: Routledge, 2018. | Includes bibliographical references and index.
Identifiers: LCCN 2017028277 (print) | LCCN 2017054513 (ebook) | ISBN 9781315645131 (Master ebook) | ISBN 9781317291107 (Web pdf) | ISBN 9781317291091 (epub3) | ISBN 9781317291084 (Mobipocket) | ISBN 9781138184503 (hardback: alk. paper)
Subjects: LCSH: Financial crises—Europe, Southern. | Europe, Southern—Economic conditions—Regional disparities.
Classification: LCC HB3722 (ebook) | LCC HB3722 .H325 2018 (print) | DDC 330.94—dc23
LC record available at https://lccn.loc.gov/2017028277

ISBN: 978-1-138-18450-3 (hbk)
ISBN: 978-0-367-36013-9 (pbk)
ISBN: 978-1-315-64513-1 (ebk)

Typeset in Times New Roman
by codeMantra

"The bread we eat in this world,
the water we drink
the strength of our knees,
the taste of our skin
is for resistance (contest)
not for our escape..."

By the Turkish Poet Turgut Uyar, (1927–1985), re-written on the walls in Gezi Park during its occupation in 2013.
Translated by Pelin Tan

Contents

Figures

Tables

Boxes

Preface and acknowledgments

Ten years have passed since the first signs of the crisis in Southern Europe emerged, and several parameters are by now well known. Dozens of books and hundreds of papers have been written using mainly macroeconomic and financial explanations. So, is it worth another book? Although I don't deny the importance of macroeconomics, in this book I shall shift the emphasis to a missing, or not well-studied, parameter, namely uneven geographical development and its relation to financialisation. I challenge the dominant view that debt, public and private, is the sole cause of the crisis. On the contrary, I argue it is the outcome of the longue durée of uneven capitalist development, not the single reason for the current turmoil in the (EU) European Union.

What happens in my country, Greece, makes me angry, and I have the same feelings for what happens in other countries in Europe, especially Southern Europe. I am aware that victims of imperialism, neoliberalism and austerity and the tragic waste of human lives exist everywhere and perhaps to a greater degree. We remember what happened a few years back in the Balkans and now in the wars in Syria, Iraq, Afghanistan and several countries of Africa. These wars are responsible for thousands of victims and refugees – men, women and children – losing their lives, most recently in the Mediterranean while escaping to Europe through its southern shores. This book, however, is about the victims of another war – this time without military acts but with the weapons of world finance and the imposition of ultra-austerity policies and dispossessions that have similar outrageous effects: deaths, emigration, hungry people and lost sovereignty in an otherwise "peaceful" situation. This book is about the crisis in Southern Europe and my frustration with what happens there, being myself a Southern European and right in the midst of it. I know that in writing a book, anger is a bad companion, and I feel obliged to warn the reader about the strength of my viewpoint and the terms under which I write.

Crises historically help capitalism to reproduce and reconstitute itself and to open new avenues for capital accumulation. Hence, the mantra "crises create opportunities" is constantly repeated without asking for whom and at what cost. Crises, however, also change the way we see things, modifying our thoughts and understanding of the world around us. The latter is consistent with the ancient Greek noun κρίση *(krisi)* and the verb κρίνω *(krino),* which mean, in both Ancient and Modern Greek, the ability to judge, to appreciate, to assess and to rethink. The Modern Latin *crisis* comes from the Greek noun κρίση *(krisi)*, but crisis as we know and use the word today has lost the meaning of the original. The book at hand uses both Greek and Modern Latin meanings. It tries to judge and rethink current crises in Southern European economies and societies in a different way, putting uneven geographical capitalist development at the centre of its analysis.

In doing this, I had invaluable help and encouragement from many friends, colleagues and comrades. The list begins with Dina Vaiou and Ray Hudson. I am grateful to both of them for reading and commenting on most of the chapters and making constructive suggestions. I am sure that they feel relief, as I do, at seeing the project completed. I benefited a lot from discussions and comments in various places and on different sections of the book, as well as from advice and data provisions from many people. So, let me record my thanks to the following: Abel Albet, Marco Armiero, Bernd Bellina, Nuria Benach, Marina Bertoncin, Rita Calvario, Myrto Hatzimichali, Evangelia Chatzikonstantinou, Giacomo D' Alisa, David Featherstone, Soledad Garcia Cabeza, Haris Golemis, Ares Kalandides, Haris Konstantatos, Maria Kousis, Olga Lafazani, Petro Marques, Enzo Mingione, Akis Papataxiarchis, Diane Perrons, Andrew Sayer, Dimitra Siatitsa, Sofia Skordili, Céline Spieker, Pelin Tan, Fereniki Vatavali and Antoni Ybarra.

Particular thanks also to David Harvey for his continuous radical inspiration and for his encouragement to finish writing by kindly asking on several of his visits to Athens and Syros "when is it coming out?" In my journey, I have missed, however, two other close friends and teachers: Ed Soja and Doreen Massey. Both of them, although for different reasons, have been "there", influencing my thinking.

I benefitted from comments and debates with participants in several conferences and seminars across Europe and the USA that have helped sharpen the argument, and from the technical assistance of my research students Thanasis Lagridis and Sotiris Koskoletos. My English editor in Greece, Robiyan Easty, corrected my "Greeklish" with patience, while also pointing out several inconsistences in the text.

Many thanks also to the following cartoonists and photographers for permission to use their art work: T. Anastasiou, C. Chappate, Coco Huang, B. Guble, Fotos movimento/Mónica Parra, Y. Ioannou, A. Petroulakis, C. Ritzler of BILD newspaper, D. Trumble and W. Warren. Also to the following newspapers: *Avgi, BILD, Kathimerini, Globe and Mail, Efimerida ton Syntakton and New York Times.*

Finally, I could not write this book without intellectual inspiration and encouragement – although they are not aware of it – from a large group of radical people in Greece, in other European countries and elsewhere with whom I feel politically close. And from a smaller, but very dynamic, group of radical students whom I have the privilege to work with – to this last group, the book is dedicated.

1 Introduction

In October 2013, I gave a talk in Brussels, invited by the (EU) European Union Directorate General for Regional Development. The session theme was "Spatial Justice in Future European Regional Development" and was organised by the Regional Studies Association, Open Universities Sessions. At that time, the crisis in Greece and in Southern Europe (hereafter SE) was fully developed; the harsh austerity policies of the Troika (the (IMF) International Monetary Fund, the EU and the (ECB) European Central Bank) since 2010 didn't provide positive macroeconomic results, while the negative effects on people were noticeable everywhere. Skyrocketing unemployment – particularly among the youth and women – salary and pension cuts, destruction of the welfare system, dispossessions of public property plus arrogance and hypocrisy among domestic and European elites made my SE fellow citizens and myself indignant. Feelings of injustice and despair fuelled demonstrations, general strikes and square occupations, making headlines around the world.

At that period, I was working on socio-spatial justice in Europe, based on my view that uneven geographical development lies behind the crisis in the EU and the Eurozone. I argued that debt, which everyone believed caused the crisis, is but one of the effects of the crisis and not its main cause. The longue durée of uneven geographical development in Europe and the uneven and undemocratic Eurozone structure was the fragile and explosive background upon which the global financial crisis was grounded and hit its first weak link, Greece. I wrote my paper along these lines, and as you would expect, there was deafening silence in the room. No one asked anything or commented. During the break, however, I had an intense conversation with three colleagues. In short, their argument was that my view is typical of Southern people's behaviour, always blaming others instead of analysing their own mistakes. I replied that this is often true, but the same can be said for EU politicians and the Troika who blame only SE instead of giving equivalent attention to their policies, to the unequal intra-European trade and the problematic Eurozone architecture, not to speak of the role of major European banks holding southern state bonds. My colleagues became more aggressive. "But we rescued you, we gave you a lot of our taxpayer's money and you sound so ungrateful", said one German regional economist.

And another, from Finland I think, added "... but anyway, you have to pay your debt back". "As we do", added the third one, from Spain. I felt I was in a tight corner, but I replied that Greek and other SE debts, particularly the Italian debt, are unsustainable, and without a political decision to cut at least part of them, they could never be repaid ... as the allies did with Germany in 1953. "But you are obliged to pay back our money; you are guilty", replied the German and Finnish colleagues, in angry tones. At that point, I realized that their statements were moving away from regional science and economic geography to become moral judgements. Although I had heard similar moral statements about the debt crisis before, in this particular case, they were, I felt, bereft of justifying arguments. The end of the break rescued me, but I was really annoyed and offended. Later, at that meeting, I decided that a reply with more complete/detailed arguments was needed, and this book is a partial outcome.

Setting the scene

> ... The blue sea, exotic beaches, bright Aegean light, hospitable locals, islands (...) make Greece an ideal holiday location. It's a country drenched in sunlight, living by the sea the year round!
>
> Greek tourist advertisement, May 2016

There is a lot of sun and bright light in SE so that everything seems all too clear. For thousands of tourists coming down Park Guëll in Barcelona, navigating upstream on the Douro, getting lost in the medieval streets of San Gimignano or relaxing on one of Naxos' "exotic beaches", the crisis in SE that everybody has been speaking about since 2010 seems invisible. Particularly on summer nights when you see all bars and coffee shops full of young people, locals included, you may ask, "Where is the crisis?" This was the reply by Silvio Berlusconi, the former Italian Prime Minister, when he was asked whether Italy was in crisis, an argument repeated by other European politicians.

However, bright light and bodily performances in tourist places, as in the advertisement for Greece, is poor evidence for understanding the underlying local social processes. Seasonality, labour conditions and low pay in these tourist places are the worst in Europe and remain invisible or of no interest to tourists. Away from tourist areas, in large cities and rural areas of the interior, ultra-austerity measures imposed since 2010 have had extreme negative economic, social and environmental effects. Austerity was the price paid by local societies in exchange for billions in financial support by the Troika to Greece and Portugal to pay back their national debts to foreign banks. Spain and Italy also imposed ultra-austerity programmes to avoid their inclusion in similar programmes.[1] By 2014, Greece had lost 15 per cent of its (GDP) Gross Domestic Product in four years, unemployment had risen to 27.5 per cent (58.8 per cent for young people), 34.6 per cent of the population were at risk of poverty, one-third was cut off from public health and the government was forced to sell off public land and public utility companies. Greece perhaps led the way in SE. Nevertheless, similar figures can be observed in the three other countries as well. By 2014, unemployment was 26.2 per cent in Spain (55.5 per cent for the young), 17.0 per cent in Portugal (37.7 per cent) and 13.5 per cent in Italy (40 per cent). In Spain 27.3 per cent of the population was at risk of poverty by 2014, in Portugal 25.3 per cent and in Italy 28.4 per cent. These are sad figures and show how ridiculous and insane the so-called "recovery plans" and "structural reforms" were, based on austerity and the "shock doctrine", wasting the most important asset – people and, particularly, the youth and women. Not all people, however, suffered during the years of austerity. A minority of the richest 10 per cent increased their wealth compared to the rest of the adult population. Following a 2016 report by Credit Suisse,[2] in 2010 the richest 10 per cent of Greeks owned 38.8 per cent of the national wealth, while by 2016 the 10 per cent had increased its wealth to 54.0 per cent. Similar figures exist for other SE countries; austerity is a class warfare indeed and the rich class that is the aggressor is winning.

Parallel to economic and political wars, ideological wars and geographical imaginations supported arguments for "lazy Greeks", "irresponsible Portuguese", "all the time partying Spaniards" and "dolce vita enjoying Italians" (see Figure 1.1).

Figure 1.1 Italian businessman exits a "bunga-bunga party" (Berlusconi's term) and one reads that Italy is now in crisis.

Source: Brian Gable from *Globe and Mail,* July 2011.

Here, it is particularly helpful to remember Antonio Gramsci and his writings about Southern Italy:

> ...(for) the propagandists of the bourgeoisie...if the South is backward, the fault does not lie with the capitalist system or any other historical cause, but with Nature, which made the Southerners lazy, incapable, criminal and barbaric.
>
> Antonio Gramsci, "Some Aspects of the Southern Question", in: *Selections from Political Writings*, 1926/1978: 444

A century passed after Gramsci wrote the above quote about Southern Italy, describing how the press, novels, articles and other dominant ideological institutions represent Southern people as "biologically inferior beings". Today, a critique of arguments naturalizing uneven socio-spatial relations in capitalism seems timely for the entire SE. In fact, from the beginning of the crisis, the dominant narrative used by European and domestic leaders focussed exclusively on SE internal inefficiencies and their tricks to fool the state, using a more nuanced language. Popularised by mainstream media and spread far and wide, the revival of old prejudices paved the way for persuading political audiences in Europe that austerity policies were what Southern people deserved and what should be used by central-northern countries to discipline them. Responsibility for the loss of economic dynamism in SE lay with the unions who strike, the inefficient state apparatus, the pensioners with high pensions, the young people who demand higher education, the universities, the political parties... We are all guilty.

I am not denying the existence of some truth in these arguments, and I do not intend to ignore the many deficiencies and corruption of SE societies and governments.[3] In the following pages, you will read many detailed descriptions of such instances. However, I would like to stress from the beginning a major concern: What happens in a particular place and society depends always on the contradictory articulation between internal/endogenous and external/exogenous conditions. Cities, regions and countries are not closed, bounded entities but are open and porous; their firms, people and institutions interact, building relations at multiple scales, from local to global and vice versa. European unification and the Eurozone intensified openness with the free circulation of capital but less so of labour. There is a shifting importance between these conditions, which are always uneven and combined in particular places and times. But uneven capitalist development is never exclusively endogenous or exogenous, because capital accumulation presupposes value circulation across space at the highest speed possible. Political and economic elites always advertise their successful policies during phases of prosperity, but try to deny their responsibility in cases of crisis, using internal or external causes as explanations. In the current case of SE, the invocation by capital and politicians, domestic and international alike, of the responsibility of entire Southern societies conceals their own "grey" role. The uneven socio-spatial structure of the EU and the unequal terms of trade and integration into the Eurozone, which meant the accumulation of debts in the South and surpluses in the North, became simply outrageous (Hadjimichalis, 2011; Lapavitsas et al., 2012; Bellina, 2013; A. Smith, 2013). Thus, national governments in Central-Northern Europe, the mainstream media and the Troika became across-the-board accusers and started blaming the victims for their debts on the basis that attack is the best form of defence.

Since then, the creditor–debtor moral obligation has been transformed into a mechanism of government with particular means of controlling political and social behaviour in SE. Debt creation under neoliberalism became the prime power relation between creditors and debtors without questioning how and by whom the debt appeared in the first place and whether it is sustainable. Thus, according to Maurizio Lazzarato (2012), debt functions: "...as a mechanism for the production and 'government' of collective and individual subjects" (p. 29), and, Lazzarato continues:

> ...it takes a particular form of *homo economicus,* the "indebted man". The creditor-debtor relationship encompasses capital/labor, welfare state services/users and business/consumer relations, just at it cuts through them, instituting users, workers and consumers as "debtors".
> (p. 30, emphasis in the original)

The indebted Man, or "homo debtor", according to Lazzarato (2012), now occupies the entire public space. Within the power relationship of debtor–creditor, Lazzarato argues, "homo debtor" is free insofar as his way of life is compatible with reimbursement. Entire societies and whole populations become indebted, producing collective subjugation, "populous debtor". Historically, debt has been used by colonising countries to exert control, and there has always been a connection between debt and guilt, as David Graeber (2011) argues, adding the dimension of morality to the uneven power relations. This is especially true in the German culture where the word for "debt" – *schuld* – famously also means "guilt". Debt defines the limits of actions and political behaviour by political leaders and people, not the other way around. For example, how "free" are indebted countries to choose a new government in elections? Direct anti-democratic interventions by the EU in Greece and Italy – where elected governments have been overthrown to be replaced by bankers – shows the new power of debt morality and resembles Foucault's definition: an action from a distance carried out on another action, which keeps entire societies "free" but disciplined and loyal to creditors.

One among the many myths about the crisis in SE is that it all started in 2008 to 2009 without any reference to the past. It is the direct effect of the

US mortgage crisis, which, by crossing the Atlantic, became a sovereign debt crisis. Until this period, European economies operated normally without problems: thus, it is clear that the crisis was due to external reasons. These narratives of the crisis succumb to what Guldi and Armitage (2014) call "short-termism", the preferred framework of economists being the immediate past, say 15–20 years, ignoring conditions present over longer periods. The Short Past has become the dogma in postmodern neoliberal economics and the framework for explanations of and policies dealing with the current crisis. This is also evident in the term "crisis" in bourgeois economics that signals a situation of disruption, a break from a prior situation of normality or stability. It demands immediate action to avoid some kind of danger, to return to normality. In capitalism, it means that some unforeseen, destructive feature inserts itself into a supposedly "normally functioning" period. Crises, however, in Marxist political economy are essential to the reproduction of capitalism and, through "creative destructions", open new possibilities and spaces for further capital accumulation and expansion. From 1970 until 2000, we had 23 major global economic crises, and from 2000 until 2015, we had one every year (see www.caprasia.com). Hence, one may ask: When was capitalism *not* in crisis? Dominant answers to this question take crisis-free capitalist reproduction for granted, and when crisis happens, it is explained without any reference to the past because it is a "radical break". Marx, however, reminds us that no social change occurs, for the better or the worse, which is not already latent within pre-existing conditions.

No doubt something very dramatic indeed happened in the EU and in the Eurozone during the period from 2007–2010. The first phase of the crisis was in August 2007 when the largest French bank, BNP Paribas, declared the closure of three of its investment funds because of the inability to finance them. The second phase started in mid-2007 with the collapse of the US housing market and the wave of foreclosures – a very material-geographical starting point indeed, but its effects spread quickly around the world to become a global crisis with the Lehman Brothers credit crunch in September 2008. The European Council, in October of the same year, promised financial support to those banks facing overexposure problems due to non-performing loans. The global crisis initially hit two weak interrelated European sectors: banks and real estate. The first signs were noticed in Spain's real estate sector (particularly housing and tourism real estate), in the former communist countries of Eastern Europe and in the Irish banking sector. To this should be added Iceland's bankrupt financial sector, a non-EU country but with many financial ties with EU banks.

The third phase of the crisis started in November 2009, when Greece became the new crisis epicentre of global capitalism, attracting headlines all over the world. With a huge public deficit of 12.7 per cent of GDP (October 2009) and an equally huge public debt of almost 300 billion euros, or around 113 per cent of GDP, it is no surprise that the country has been into the "sausage grinder of the financial markets and international banks" (Golemis, 2010: 129) as they are the principal holders of Greek state bonds. At the same time, an aggressive attack by international financial speculators began, increasing the cost of borrowing from international institutions. The immediate trigger in the Greek case came in February 2010 when the so-called "socialist" government of (PASOK) the Panhellenic Socialist Movement was forced to seek refinancing for more than 60 billion euros of debt. During the heady days of irrational financial speculation of the late 20th and early 21st centuries, global banking institutions "conspired" with the Greek government to conceal the extent of the debt holdings of the country.

Box 1.1 Greece's secret deal

The journal *Spiegel*, usually echoing the position of the German government, in its electronic version (2 August 2010), describes how Greece's debt managers agreed to a huge deal with the savvy bankers of the US investment bank Goldman Sachs at the start of 2002. The deal involved so-called "cross-currency" swaps in which government debt issued in dollars and yen was swapped for euro debt for a certain period to be exchanged back into the original currencies at a later date. Such transactions are part of "normal" and "legitimate" government refinancing, uncontrolled by Maastricht rules. But in the Greek case – as happened with Italy and even France in previous years – the US bankers devised a special kind of swap with fictional exchange rates. This enabled Greece to receive a far higher sum than the actual euro market value of 10 billion dollars or yen and thus to enter the Eurozone with a 1.2 per cent of GDP deficit. In that way, Goldman Sachs secretly arranged additional credit up to 1 billion dollars for the Greeks.

Due to the specific organisation of the Eurozone and the inability of ECB to act as the "lender of last resort", the crisis caused a sobering up, and skittish investors downgraded Greece's risk rating and withdrew from further refinancing, causing the February crisis and finally the intervention by the Troika. Then it was Ireland in 2010 and Portugal in 2011 who, under severe pressure, asked for similar intervention by the Troika. Spain and Italy, having similar high debt levels, avoided the Troika's intervention but committed themselves to hard austerity measures and fiscal reform from 2011 onwards. Spain, with the help of the ECB, created a 99 billion euro bailout fund to rescue its vulnerable banks, shifting the burden to pension and salary cuts. Finally, Cyprus in 2012 also asked for the Troika's support for its banks. Different economic conditions and political decisions lie behind each case. However, as all governments moved in to bailout banks and financial institutions with taxpayers' money, public debt became a common problem to all, together with cuts in public spending, introducing deep recession and widespread poverty.

Why did that happen? Undoubtedly, the global financial crisis, as a conjuncture, played a decisive role in revealing every country's weak points. But an explanation focussing exclusively on this "break" and the rise of debt, public and private, seems less convincing. My supplementary hypothesis contains three elements. First, the roots of the crisis are embedded in uneven geographical/regional development, which characterised the socio-spatial structure of the EU long before the crisis, at least since the 1980s, combining endogenous and exogenous elements. Public debt is not the sole cause of the current crisis but one among many of its manifestations, while its roots are located in the gradual decline of southern regions' productive and export performance. Second, SE economies participated, to varying degrees, in a major capitalist transformation characterised by a shift from productive to rent-seeking activities, from the primary to the secondary circuit of capital. This transformation provides "thrive profits to rentiers while work does not pay", as Standing (2016) powerfully argues, and it is visible in most developed economies of the global North, particularly in the USA (see also Harvey, 2010; M. Hudson, 2010; Sayer, 2015), but, interestingly enough, not in Germany. A major parameter in this transformation relates to the rise of the so-called "FIRE" sectors, i.e. Finance, Insurance and Real Estate. It is not accidental that the first signs of the crisis in SE were found in sectors belonging to these "rent-seeking" activities: finance and real estate. And third, in the context of neoliberal hegemony, particular elites and regional hegemonic blocs in Central European countries (in the original EU-6 and above all in Germany) designed the Maastricht Treaty and the euro to conform to the characteristics and needs of their economies. In addition, in order to discuss the problems of the euro, they established undemocratic, non-elected institutions such as the Euro Working Group, which, together with other unaccountable EU institutions, form powerful multi-scalar lobbies to promote capital's interests. In doing this, they took advantage of the inevitable social and spatial restructuring that accompanied the introduction of the euro to regain political and economic control, not only vis-à-vis global financial markets but within the EU as well. This restructuring was based on pre-euro conditions of uneven economic and geographical development within the EU, which accelerated and intensified after the introduction of the euro to become the crisis-driven restructuring we face today.

Uneven geographical development as a framework enables us to approach the roots of the crisis in a dialectical way. Contradictory and multi-scalar processes such as fixity and motion, territorialisation and de-territorialisation, the flow of capital and information across space, the relative immobility of labour, place-specific devaluations, path dependency and institutions regulating all the above produce and reproduce uneven geographical development. The social and competitive organisational space under capitalism introduces monopolistic competition providing quasi-rents to certain sectors and locations while marginalising others. Unevenness across space and sectors is not a mere sidebar to how capitalism works but fundamental to its reproduction (Harvey, 1982, 2010; R. Hudson, 2005). Whatever it's particular source, uneven geographical development is a contributing factor to the creation and maintenance of individual and collective inequalities and hence to social and spatial injustices (Soja, 2010). To act against these injustices requires a politically sensitive approach towards development. In addition, it requires multi-scalar institutional and informal networks of solidarity, from global to national and to local/regional scales. Socio-spatial justice and socio-spatial solidarity are principles and values of progressive planning, neglected today under the neoliberalism ideology, which promotes de-politicisation. I strongly believe that we always need to talk about socio-spatial justice and socio-spatial solidarity in periods of crisis, such as the current crisis in the EU, in the context of which a discussion on theories, principles and values becomes more than imperative.

Related to social, political and spatial justice are social movements in SE cities that are resisting austerity with demonstrations, square occupations and networks of solidarity, highlighting once more that urban space and place matter. Since 2008, and particularly since 2010, there have been hundreds of bottom-up resistance and solidarity social movements all over the cities of Spain, Portugal, Italy and Greece. Building upon militant experiences of the 1990s and the 2000s, recent social movements in SE have shown that many people in these countries are active radical agents of resistance and solidarity, not passive victims of crude ultra-neoliberal policies. In addition, they are demonstrating the inadequacy of traditional progressive political forces of social democracy and the left.

The Eurozone crisis meets Antonio Gramsci

From the beginning of the crisis, EU and Eurozone non-elected institutions repeatedly showed arrogant and authoritarian characteristics, reminiscent of Nicos Poulantzas' analysis of "authoritarian statism". Elected governments were forced to step down and replaced by bankers; national Parliaments were replaced by unconstitutional agreements, so-called "Memoranda", as with the Troika. The role of the European Parliament was reduced, while the ECB's role was strengthened; private interests increasingly intruded into public decision-making; national sovereignty was equated with financial credibility, meaning that if you lose the second, you lose the first as well (*Monde-Diplomatique*, 9 December 2012). In Athens and Lisbon (and Dublin and later Nicosia), Troika's "men in black" dictated austere economic and social policies, while Madrid and Rome implemented austerity under the threat of similar interventions. In order to understand this anti-democratic turn in European politics, I found it useful to revisit Antonio Gramsci, a perceptive observer of the 1930s crisis. EU institutions were long ago accused of having permanent democratic deficits.

During the crisis, however, the EU and Troika's policies became even more authoritarian, displaying what Gramsci described for the 1930s as "caesarism", an appropriate description of European politics in the 21st century (*Monde-Diplomatique*, 9 December 2012). According to Gramsci, during capitalist crises, public institutions elected by universal voting, such as parliaments, are pushed into the background. Crisis conditions are strengthened instead:

> ...the relative power of bureaucracies (political and military), high ranking financial circles, the Church and in general all institutions which are unaccountable.
>
> ("War of movements and war of positions", *Selections from Prison Notebooks*, 1971: 345)

In the Eurozone, a case par excellence is the Eurogroup. It is an informal institution, as dogmatic and exclusive as the Church, where finance ministers from the euro countries – quite often without an economics degree – participate "as friends", with an appointed president, without any democratic accountability and without official minutes, even though they dictate the conditions of everyday life for millions of Europeans. To his credit, Yanis Varoufakis, the first Syriza finance minister, exposed officially the undemocratic procedures of this peculiar and exclusive "Euro club" in several of his interviews. The way the Eurogroup reacted, together with other unaccountable institutions such as the ECB and EU bureaucrats, against the Greek referendum in July 2015, against the centre-left Portuguese government in 2016 and against the British Brexit in June of the same year, is indicative of this "ceasarism".

Gramsci, a prominent Italian Marxist and founder of the Italian Communist Party (Partito Comunista Italiano, PCI), studied in depth the unification of the Italian state, the Risorgimento. He analysed how the unevenly developed, economically and culturally, Italian regions were unified "from above", without any democratic popular participation, under the dominance of the northern bourgeoisie in alliance with Southern landlords (latifundisti) (Nardone, 1971). Although "history does not repeat itself" and major differences exist between the Risorgimento and the construction of the EU and the Eurozone, both processes have a common characteristic: capital and particular elites guiding both with the total exclusion of the popular masses. Gramsci is also known as a spatial theorist for taking account of the particular geographical embeddedness of historical phenomena (Jessop, 2005a; Ekers et al., 2013; Featherstone, 2013). For him, space and place are related with everyday life, with common sense and with collective memory, arguing that social and cultural identities have a spatio-temporal depth (Vaiou and Hadjimichalis, 2012). In this context, he made interesting observations on the conditions of uneven development between Northern and Southern Italy in various writings and especially in his essay "Some Aspects of the Southern Question" (in: *Selections from Political Writings* (1921–1926/1978). Central to Gramsci's views of uneven development, apart from capital's expansion, is the role of the state during the Risorgimento and the cultural and political unevenness between Northern and Southern Italy (the Mezzogiorno). Gramsci didn't have a theory of uneven development as such but a vantage point from which to theorise about socio-spatial inequalities, which could turn against the weakest part if counter-active policies were not enacted. He gives an example describing the articulation between the fiscal and customs systems of the new Italian state, with mobile capital making localised accumulation in Southern Italy impossible "on the spot":

> ...Any accumulation of capital on the spot, any accumulation of savings, is made impossible by the fiscal and customs system, and by the fact that the capitalists who own shares do not transform their profits into new capital on the spot, because they are not from that spot
>
> (*Some Aspects of the Southern Question*, p. 16, op. cit.)

These observations at the beginning of the 20th century are timely today, while keeping in mind major differences between then and now. They constitute useful starting points to analyse the current crisis in the Eurozone, and I note four of them below. First, Gramsci's attention not only to capital accumulation but also to the articulation between capital accumulation, politics, political parties, cultural differences, institutions and above all to the role of the state provides a useful framework to understand the current crisis in the EU and the Eurozone. Here, particularly useful are his points on "caesarism" during crises and the strengthening of unaccountable institutions. Second, although he is a prime analyst of the conjuncture, in doing this he always takes into account the historical and socio-spatial roots of the conjuncture. Third, his attention to specific cultural and economic characteristics, to class alliances and to the role of regional hegemonic blocks enables him to understand the difficulties in building solidarity across boundaries among working people. But he insists on the necessity of solidarity among culturally and geographically different subaltern classes. This point is particularly useful in understanding the sporadic expressions of solidarity towards SE people by others during the crisis in Europe, with the notable exception of those belonging to the left, to anarchism and to some organised trade unions. At the same time, Gramsci provides the basis for opposing prejudices used by both sides, Northern Italians accusing Southerners and vice versa. And fourth, uneven development between North and South in Italy is analysed using capital accumulation processes but also by going beyond them, with an analysis of different subaltern geographies of connection between the South and North, particularly the role of migration.

This book

My analysis focusses on Portugal, Spain, Italy and Greece. When I refer to them, I am aware that I don't include the complete geographical area of SE, but my choice is based on the fact that these four belong to the Eurozone, and they have faced severe crises since 2009–2010. The four countries and their regions share significant dissimilarities as well as similarities, Italy being the third EU economy (after Brexit) and Spain the fourth, while Greece accounts for only 2 per cent of EU GDP. But, as I explain in the following chapters, there are other good reasons to study them together. Throughout the book, I give particular attention to politics and particularly to their spatialisation and scale as important regulatory frameworks. Due to particular SE traditions, national and regional political parties, from the left to the extreme right, have played key roles, and this is true for local bottom-up social movements. I give attention to both because they do politics at different scales, and the scalar approach helps us to understand the labyrinth of SE politics.

In Chapter 2, I deal with uneven development before the euro, from the 1980s to the late 1990s. I give particular attention to several common characteristics in SE, such as the dominance of small and medium enterprises (SMEs), the informal sector, the fragmentation of regional labour markets, the role of the family and the role of the weak, clientelistic and familistic state. These characteristics allowed the rise of what I call dynamic "intermediate" regions that faced severe problems in the 1990s after crucial changes in the European and global division of labour. The Maastricht Treaty was a major turning point. It institutionalised neoliberalism and by transferring major regulatory powers to EU bodies it badly affected southern states, gradually destroying their national financial systems and their regional productive structures. Thus, SE regions and states, with variegated economic and institutional dynamism among them and more so compared with the central-northern areas, entered the Eurozone ill-prepared, deepening the intra- and inter-regional unevenness.

Chapter 3 discusses capitalist transformation and the building of the Eurozone. Here I challenge the dominant view that debt, public and private, is the sole cause of the crisis, arguing that it is the combined outcome of financialisation and the longue durée of uneven capitalist development. To support my argument, I look first at deeper capitalist transformations towards financialisation, particularly investment in assets and profiting from different types of rents, thus making a switch from the primary to the secondary circuits of capital. I focus especially on real estate, a geographical sector par excellence, and its boom-bust cycle that preceded the Eurozone turmoil. Then, I discuss the formation of the Eurozone, highlighting the absence of a geographical and regional perspective in the so-called "national convergence criteria" and argue that uneven development and uneven trade flows, not only debt, tell the story of the crisis. At the end of the chapter, I discuss the hybrid space/scale of the Eurozone and the role of undemocratic multi-scalar governance that fuelled the effects of the crisis and made it impossible for EU elites to solve any problem that had not been foreseen.

In Chapter 4, I shift the emphasis to ideological wars, to imagining and constructing the New "Southern Question". From the beginning of the crisis, dominant elites and the mainstream media constructed a narrative that framed Southerners who "live beyond their means" and "their" governments who systematically cheat EU institutions as the sole cause of the crisis. Starting with the famous "PIGS in Muck" story, I discuss geographical imaginations and prejudices about the South that have a long history but remained dormant during the years of prosperity. I discuss the geographical imagination and historical determinism of Robert Kaplan, a senior fellow at the Center for a New American Security, his static, deterministic and essentialist view of geography and history. In the last section, I present a short overview of prejudices and stereotypes used by politicians and the mainstream media against SE people and governments, as they have been published in newspaper articles and in cartoons.

Chapter 5 deals with the problem of the de-politicisation of uneven development, focussing on my field, economic geography and regional and urban planning. I argue that the gradual shift of major theories of economic geography and regional development towards "Third Way" thinking became a new de-politicised orthodoxy at a time when neoliberalism made a frontal attack in the field. This has made it easy to absorb their views into neoliberal policies, and when the crisis began, there was only one paradigm on the table – the neoliberal one. In this respect, both before and after the crisis, EU policies related to regional and urban inequalities were practically absent, other than for the rhetoric of "social and territorial cohesion". Thus socio-spatial inequalities increased dramatically after the crisis, resulting in severe injustices. The unfair terrain of the Eurozone and socio-spatial injustices among states and regions are further discussed using particular examples and data on unemployment, the risk of poverty, severe material deprivation, brain drain and energy misery.

Chapter 6 enters the terrain of resistance and solidarity in SE. It focusses on demonstrations, squares occupations and solidarity social movements in urban areas that resisted harsh austerity policies. These movements challenge de-politicisation as the main characteristic of neoliberal austerity via the spatialisation of democratic policies. The chapter makes a link between past and present SE militant left traditions and analyses the new political subjectivities that came out of mass demonstrations and the occupation of squares. These include Rossio in Lisbon, Puerta del Sol in Madrid, Plaza Cataluña in Barcelona and Syntagma Square in Athens. Finally, it discusses the extensive network of bottom-up, self-organised solidarity movements that spread across SE to provide urgent assistance to those in need, focussing on three Greek paradigmatic examples: the provision of food by the "food without middlemen" movement, the health solidarity services of self-organised "social clinics" and solidarity movements supporting migrants and refugees.

Finally, Chapter 7 discusses politics and poses questions about the future of the EU and the Eurozone. Are there any politics of hope, or do we face the time of monsters, as Gramsci predicted from his fascist prison in Italy before the Second World War? The chapter discusses developments after the

60th "birthday" celebration of the EU in Rome. It criticises the "multiple speed" proposal as the new European paradigm that rationalises and institutionalises the very causes of the crisis, namely uneven geographical development. At the end, I offer some thoughts on possible politics of hope, based on my own SE experience, imagining a kind of anti-capitalist utopian pragmatism.

Writing this book I several times faced the question of whether surveillance and actions by the Troika are legal and legitimate. The short answer is a strong *no*, also supported by decisions of supreme courts, which declared as illegal many policies imposed in cases such as lay-offs, pension cuts and dispossessions. Greece is again the extreme case, still being under continuous surveillance and control at the time of finishing this book. Portugal remained under similar conditions until 2014, but the EU and ECB continue all kinds of "interventions", as they do in Spain and Italy, particularly the tough surveillance of their banks and national budgets. The latter is the application of the new, undemocratic, economic governance of the EU, established in 2010, after the Greek crisis. All these actions support the idea that world finance used SE as a mode of warfare in developed economies, convincing the governments in the four countries to pursue policies that destroyed, to different degrees, their own economies. Giorgio Agamben (1998) calls forth the notion of "condition of exception" as a governance model based on the distinction between the juridical order over a particular place and another order in which people are deprived of similar juridical protection. In the absence of a unified juridical-political space, like the EU, dominant elites could exercise the particular spacing of exception by declaring a state of emergency due to the ideology of "crisis as break", as happened with the Troika and the Memoranda. In this situation, what is at stake is not only what actions and policies are considered legal and legitimate but also what lives or forms of lives count. In other words, do we accept a canonical process of life management, an ordering of space, that allows, as Gregory (2004) suggested, conquest "without taking the land", although the latter can be materialized by dispossessions? Do we accept that the Troika and other EU institutions are able to intervene in everyday political decisions from a distance, using debt as a battering ram?

Notes

1 We need to remember that Hungary, Latvia and Romania were the first countries to apply the painful IMF and EU programmes to stabilise the exchange rate. Austerity policies resulted in the well-known devastating effects: deep recession, high unemployment and severe cuts in social protection.

2 The Credit Suisse report defines as net wealth the value of financial assets plus real estate assets owned by households minus debts. Private pension fund assets are included, but not entitlements to state pensions.

3 Notwithstanding the complicity of big German corporations in the biggest scandals in Greece, for example, the legality of bribes by German companies, only recently revoked, and the huge amounts lost to fraudulent welfare payments in the UK.

2 Uneven development I

Capital restructuring and changes in the spatial division of labour before the euro

(SE) Southern European uneven social and productive structures pose serious difficulties for a comparative analysis. In order to interpret data on different temporal and spatial scales, and from different sources, and to answer macro-quantitative and micro-qualitative questions, I rely on the hypothesis that southern countries and regions constitute variants of a particular capitalist development path, which takes into account important national and regional variations (Hadjimichalis, 1987; Mingione, 1991; Bruff, 2011). This path is equivalent neither to the North-Central European industrialisation model, nor to the peripheral underdevelopment model, as known in former colonies. Arrighi (1985, 1990) and Wallerstein (1983) use the concept of *semi-peripheral capitalism* to highlight specific processes of institutional change and economic difficulties in SE.

The issue of whether we can generalise – or at least, the extent to which we can generalise – about an SE model, is an open one. The history of development in the South, however, does not conform – with notable regional exceptions in Northern Italy, Barcelona, Bilbao and Lisbon, among others – to the description of the Fordist factory, the mass worker and the welfare state. Its capitalist development path is different but this does not mean "less important", "backwards" or "less developed" (Mingione, 1998a, 2000). Dominant discourse on development, coming mainly from Anglophonic and Germanic countries, characterises experience of very limited areas in the world as global and undervalues the development histories of the rest. SE is no exception, and the lack of appreciation of difference is part of today's problems, initiated during the process of the EU integration and particularly for countries that adopted the euro.

Developments in the 1980s

The political and social situation in SE at the beginning of the 1980s was dramatically different from that of the previous decade. Long-established dictatorships in Spain and Portugal and the more recent one in Greece had been overthrown; political parties had been legalised, political prisoners were released and people in forced exile returned home. Furthermore, in 1981, in Greece, PASOK won the national elections and formed the first-ever socialist government. By 1983, in Italy, (PSI) Partito Socialista Italiano had made a five-party coalition government, which lasted until 1986. In 1982, in Spain, (PSOE) Partido Socialista Obrero Español gained a great victory in national elections. In Portugal, the (PS) Partito Socialista and the (PSD) Partito Social Democratico made a coalition government from 1983 to 1985, and thereafter PSD remained in power alone throughout the 1980s and into the first half of the 1990s.

In terms of relations to Europe, during the same decade, Greece joined the EU in 1981 followed by Spain and Portugal in 1986. Accession to the EU had major political and economic consequences, which were not realised in the early years of membership.

Box 2.1 A socialist Southern Europe?

Although the declaration by F. Mitterrand and A. Papandreou in 1981 of a two-tiered Europe "conservative in the north and socialist in its southern periphery" was an exaggeration, politics in SE was indeed more to the left, or more progressive, compared to the rest of Europe. Because apart from having socialist-social democratic parties in power, the countries in SE also had strong Communist parties in regional and urban governments – the Italian PCI was the largest and most influential – plus strong and militant unions and many far-left small parties, anarchists and militant groups. The latter were not able to enter national parliaments, with the exception of Italy, but were quite influential among intellectuals, students, academics and social movements. These developments, however, did not form a "left hegemony" in Gramscian terms, because right-wing conservative parties continued to be strong while the presence and strong influence of the Catholic Church in Portugal, Spain and Italy and the Orthodox Church in Greece, supported conservatism. Nevertheless, with important variations among the four countries, policies of redistribution took place throughout the 1980s. Notable were real wages increases, universal health care, the rise of social security contributions, free access to higher education and other fringe benefits. Parallel to these improvements, the public sector remained highly inefficient, tax evasion continued and all governments were forced to borrow money from international markets.

After the end of the 1970s, capitalist development started to flourish, not around the new industrial growth poles and port-industrial complexes[1] as a planned trickling-down effect, but spontaneously in other regions and localities whose economic performance, social division of labour and degree of state intervention was at an *intermediate level*. The pattern has become clear in Italy since the late 1970s, in Spain since the early 1980s and in Greece and Portugal since the mid-1980s (Hadjimichalis and Papamichos, 1990) and combines both internal/endogenous and external/exogenous conditions at the national and global level.

The new pattern did not challenge older industrial and urban centres such as the golden triangle in northern Italy (Torino–Milan–Genova) nor the importance of major urban regions such as Athens, Rome, Madrid, Barcelona, Bilbao, Lisbon, Oporto and Thessaloniki. Neither did it seriously impact established centres of mass tourism along the coasts and the islands such as the Costa del Sol, the Balearic Islands, the Algarve, the Adriatic coast, Corfu, Crete and Rhodes. The bulk of economic activity is still to be found there, while foreign direct investment in large industrial branch plants, such as the Ford plant in Valencia and the Opel plant in Zaragoza, preferred green field regions. Instead, the new pattern highlighted the expansion of small-scale manufacturing activities in non-metropolitan areas, the extensive transformation of agriculture and the restructuring of services and real estate in areas without prior major dynamism.

In retrospect, by looking at these developments, (see also Paci, 1972; Bagnasco, 1977; Vàzquez-Barquero, 1992) I can identify five major restructuring issues more or less common to the four countries, bearing in mind the obvious: SE is far from a homogeneous entity; it shares some common characteristics but also important differences (Hudson and Lewis, 1985).

First, in the 1980s *SE regions entered the international and particularly the European division of labour as preferential partners of North-Central economies* and especially in the case of Spain for (FDIs) Foreign Direct Investments. Despite the existence of large, dynamic industrial firms in Italy and Spain, the majority of industrial enterprises were very small, with less than 10 people – a persistent structural characteristic from the 1980s onwards until today. Figures in Table 2.1 show that SMEs are widespread in the rest of the EU, but SE dominates in micro-firms and the percentage of SMEs per 1,000 people. Micro- and small firms characterise all sectors including agriculture, services, trade and tourism. Some are indeed backwards, with a short-term planning horizon, low technology and limited knowledge of the business environment. At the same time, there are thousands more that are performing the economic in innovative and novel ways and are very successful in exports to international markets.

After the crisis of Fordism, the search by capital for more flexible production systems in all sectors had social, technological and territorial consequences. It was the period during which analysts, mainly from Italy and then from Anglophone countries, discovered successful applications of flexibility in industrial districts (IDs) in "Third Italy" and, using them as a model, identified similar cases in other parts of SE.

Table 2.1 Micro, small and medium enterprises in SE and EU, 2012

	% SMEs in all sectors				
	Micro 0–9	*Small 10–49*	*Medium 50–249*	*SMEs per 1,000 people*	*Employment in SMEs as % of total*
Germany	88.3	10.2	1.5	36.4	70.4
Greece	97.5	2.1	0.3	72.2	74.0
Italy	95.6	4.0	0.4	77.8	73.0
Portugal	94.5	4.7	0.7	70.2	73.3
Spain	94.4	4.8	0.6	67.4	78.6
EU	91.8	6.9	1.10	48.6	67.0

Source: EC Enterprise and Industry (2012).

National and European regulatory policies helped the performance of small firms, such as several competitive devaluations of national currencies[2] and the introduction of the international Multi Fibre Agreement in 1974, protecting EU firms in the textile, clothing and footwear sector from imports from low wage countries. This particular specialisation in the production of consumer goods related to fashion, in agricultural high-value products and in mass tourism is strongly linked to three factors: the uncertainty of demand, taste and fashion patterns; the demand for low wages; the maturing of existing technologies that move towards important labour-savings; and more spatially diffused forms of production.

Box 2.2 Third Italy

In the late 1970s, the sociologist Arnaldo Bagnasco published his book *Tre Italie* (1977), questioning the North-South divide in Italy. He described the particular economic dynamism in Northeast Italy and particularly in Veneto, Emilia-Romagna, Tuscany and the Marche. The dynamism based itself on dense networks of small family firms in IDs, located in small historical towns, combined with agriculture and tourism. Each ID is specialised in a particular product, such as ceramic tiles, textiles, clothing, footwear, furniture, musical instruments and others, following the localities' historical specialisations. While Italian scholars were sensitive and careful in their analysis, Anglophone researchers were enthusiastic about Third Italy. They soon made it an ideal model and the "post-Fordist" alternative for stagnating industrial areas in Europe and North America. In the extended literature, the Italian districts of "Third Italy" stand as landmarks and standard points of reference. Italian IDs, together with other "new industrialised spaces" were used as paradigmatic examples of "good practice" for other regional economies in Europe and beyond.

And so, a whole series of analytical and policy-oriented concepts have been born, inspired in part or totally from Third Italy, such as "second industrial divide", "resurgent regional economies", "networked firms and regions", "learning regions", "innovative firms and regions", "endogenous development", "local development", "the embedded firm", etc., best summarised in what is commonly known as "New Regionalism" (for a critique see Blim, 1989; Hudson, 1999; McLeod, 2001; Hadjimichalis and Hudson, 2014).

Second, a specific regional characteristic that calls for attention is *the importance of the informal sector* (Capecchi and Pesce, 1983). Although informal activities are important and historically widespread in SE, or perhaps for this very reason, the situation has attracted attention only to criminal activities such as drug trafficking, prostitution, tax evasion, toxic waste dumping, etc. However, I am referring to a whole set of everyday practices that are not illegal but are culturally embedded "ways of doing or performing" the economy, with processes and practices that permeate the ensemble of social relations and activities and are not limited to criminal or illegal activities (for a detailed discussion see Vaiou and Hadjimichalis, 2004). For this term does not only include excess profit-making by breaching or avoiding the law, but also conditions of hard work and strategies of survival. It inevitably draws attention to those involved in such practices: the concrete men and women – local or immigrant – and sometimes children, who contribute to the flexibility of small firms and to the dynamism of particular places. It also draws attention to those who bear the effects of the decline of economic activities – people who are gendered and ethnicised, who work not only for the market but also within families and communities. This labour pool does not conform to descriptions of formal, full-time employment but that means that it is marginal or outside the capitalist relations of production. Informal activities are determined largely by the "formal" regulatory system (Mingione, 1995); in this context, it is not the lack of control of informal activities that calls for attention, but their specific integration/subordination in the accumulation process through specific actions (including tolerance) by different social actors and institutions.

Third, *segmentation in local/regional labour markets* is an important component of the new flexible development pattern. In many localities in which dynamic growth occurs – and with important variations from place to place – segmentation effectively combines formal and informal employment, with less unionisation. This segmentation characterises local labour markets in regions with a long industrial tradition, a substantial working class and chaotic urban growth with inadequate housing and welfare provision. Average incomes are more than double that of the rest of the country, and this helps explain continuous rural to urban migration. In mainly agricultural regions with limited tourism, segmentation derives from rapid de-ruralisation, high unemployment and rural exodus. State expenditure on public infrastructure provided employment in construction and often made local populations "victims of political patronage by state apparatuses", as Pugliese (1985) argued. In intermediate regions, social stratification and segmentation in labour markets was centred on family business, and capitalist production was directly linked to social reproduction (Mingione, 1991). The social picture is completed by mentioning the gender composition of the various groups of working people. The definition of unskilled, seasonal and low-paid labour is always related to the gender of its bearers. This has to do more with jobs being identified as "women's work" (like sewing, caring) than their technical requirements (Vaiou and Hadjimichalis, 2004).

A crucial agency in all the above is *the family* whose members (taken as a group) have multiple forms of employment. Some have full employment; others are engaged in informal, home-based activities; some work seasonally in agriculture or tourism; others remain unemployed or perform unpaid family labour. Households in SE operate as multifunctional, flexible agencies providing selective security to family members at considerable cost to those who are subordinated to the family head, usually the men. This role of the extended family is not new in SE, but adopts and capitalises on opportunities arising from the restructuring of modern economic activities. Working conditions and commitments within the family unit not only tend to increase, taking up nearly all available time, but also vary discriminatorily between genders and age groups. Old patriarchal and authoritarian relations are reproduced within the family, with women at the bottom of the hierarchy (Blim, 1990).

Fourth, *tourist and construction sectors* give rise to employment, mainly temporary and/or seasonal, and they are highly important in all SE regions, particularly in cities, coastal areas and the islands. Local people can have multiple sources of income working in construction and during summer in a small tourist firm, while they can also participate in periodic farming work (Williams, 1984). Together with work in small industrial firms, multiple employment opportunities in SE are a major source of informalisation and constitute an important social security net. However, they have attracted minor attention as a source of flexibility and dynamic growth because researchers have been focusing, on the one hand, on industrial small firms only, and on the other, never show interest in pluriactivity and the informal sector, ignoring the complementary role of all the firms in a region, large and small. Ironically, many writers and policy makers have used Third Italy and Valencia as their model, but this has not been of help in understanding how small-scale industry interacts with small-scale tourism, how pluriactivity in agriculture and construction sustains both and finally how multiple employment in these sectors was, until the late 1990s, the corner stone of family reproduction (Blim, 1989; Smith, 1999).

Finally, yet importantly, the fifth issue concerns *formal institutions, particularly the state and local authorities*, which play an important regulatory role through their active policies and their passive tolerance of the situation. After the mid-1980s, EU institutions and policies took an active role in shaping local development. Among these policies, and with variations among countries and regions, I could mention labour legislation, regional incentives, allocation of public investments, the (CAP) Common Agricultural Policy, specific projects financed by the EU and tax exemption of agricultural incomes up to a specific amount. Active policies were accompanied by inefficiency and corruption by central and regional governments, especially in land-use control and planning permissions, together with limited power to control tax evasion. In many regions (e.g. Mezzogiorno, Crete, Andalusia, the Mesetas, Peloponnese and South Portugal), economic activities depended directly and indirectly on state choices. Thus, political affiliations with political parties in power became more important than social or economic factors. Political clientelism and patronage found a very favourable environment in which to develop, particularly as they did in PASOK's Greece and PSOE's Spain (Giner and Sevilla 1984; Spourdalakis, 1988).

Although these issues have been integral to the new development pattern in SE, during the 1980s and 1990s attention by mainstream policy-makers and theorists focused only on the paradigmatic IDs of Third Italy. In doing so, they made two important mistakes. First, in their research, a very selective appropriation of the complexity and richness of Italian IDs has taken place, ignoring the wider SE development path described earlier in which paradigmatic small firms operate. Second, and as an outcome of the first mistake, only certain general economic, organisational and institutional issues have been considered, while others such as the protectionist nexus of small firms, informal work and pluriactivity, the segmentation of the labour markets, patriarchal power, gender and ethnicity, the generation gap, low wages and working conditions – to mention but a few – remain in the dark.

Post-1990 developments

The fall of the Berlin Wall in 1989 and the triumph of Western neoliberal capitalism over Eastern-state pseudo-socialism were celebrated by the dominant classes in Europe, as was Europe's dissociation from US and Soviet domination. In October 1990, the Federal Republic of Germany integrated the (DDR) Democratic Republic of Germany and formed the largest economy in the EU. This substantial change in the political and economic dynamic of the European balance of powers constitutes a major turning point, with many implications to come. Seventeen years after the fall of the

Berlin Wall, however, many hopes from that historical time have proven to be unfulfilled. Europe, incapable of understanding, acting in and solving the Bosnia situation, was forced to accept the Dayton agreement, which placed the USA again at the centre of European affairs and confirmed its dominant position. And the triumph of Western neoliberal capitalism over Eastern states, as was expected, has been simply a destruction of what existed before, with little hope of recovery. Thus, in the EU, the 1990s started with two inter-related geo-political and economically painful processes, with negative consequences for Southern and Eastern members and for the democratic process of European integration itself.

The first concerned the *civil war in ex-Yugoslavia*, which started in the early 1990s with the strong involvement of the EU, NATO and the USA. The war re-affirmed the legacy of military intervention just as the EU was attempting major internal reform and eastwards expansion, expressions of different ways of imagining European space (Chomsky, 1999; Hadjimichalis and Hudson, 2004). Serbia and Kosovo are strategic regions controlling the Danube and the exit from the Balkans to the Adriatic Sea, and thus very relevant to the European expansion project towards the east. As Tariq Ali (1999: 234) wrote: "...the need to protect the Kosovars served as the pretext for NATO's bombardment, but its real aim was to secure its control of this strategic region and fortify an extensive NATO bridgehead in the heart of the Balkans".[3] At the same time, as Pierre Bourdieu (*Humanité*, 18 May 1999) argues, it was used to "...discipline future EU members with a similar communist past in the most undemocratic way". European cities were bombed for the first time since the end of the Second World War in its eastern periphery, having been labelled as "less democratic", "culturally different" and "where human rights are violated". The construction of knowledge through such language produced meanings and geographical imaginations for the Balkans, not only of the type "unlike us", but also as a space where all negative Western characteristics of the past could be dumped. Nationalism, war, political instability and riots became identified with this particular area so that a current positive European image could be juxtaposed. What Western Europe hates most is not its image at the dawn of civilisation, but its image a few generations back. Shifts of this kind create a new type of "democratic deficit" embodied in military actions. The war in ex-Yugoslavia represents a critical step backwards in European politics, a painful reminder that European integration has a military arm in which NATO and the USA are partners.

These military acts, without UN authorisation, have had a long-standing negative effect on European politics, directly undermining social compromises and introducing military *faits accomplis*. They also undermined centre-socialist governments and coalitions at that time in power in France, Italy, Germany, Portugal, Spain and Greece. They posed, directly and brutally, the question of which political force would lead the project of European integration: conflict-centred neoliberal imperialism or Euro-stakeholder capitalism searching for negotiated solutions. The war in the Balkans constrained and disciplined the social democratic political forces then in power in many states of the EU, serving as a Foucauldian disciplinary technology, which opened the path for crude neoliberal policies and, since the mid-1990s, to right-wing governments.

The second interrelated and painful process concerns economics and politics and has a name: the *institutionalization of neoliberalism*. It acquires different applications in particular social formations, and therefore, it seems more appropriate to speak of "embedded neoliberalism" or "real existing neoliberalism" (van Apeldoorn, 2002). As a political project, it reconfigures the institutional space of capitalism and, consequently, its economic logic. At the EU level, the institutionalisation of neoliberal ideas begins with the known Stability and Growth Pact introduced in the Maastricht Treaty of 1992 for monetary union (see Box 2.3) ensuring that the dogma of fiscal stability cannot be reversed. It set policy in the interest of big business, including market deregulation and protection of the financial sector. Together with the (EMU) European Monetary Union and the euro, launched in 1999, they have introduced a rigid regulatory framework of permanent austerity and "foreign" surveillance, i.e. from Brussels and Frankfurt, over national economic policy (see also Chapter 3). The Treaty was far from an attempt to turn the EU into an

Box 2.3 The Maastricht Treaty: financial colonialism redux

The Treaty was structured on three pillars. The first pillar alludes to community dimension and comprises existing and new important institutions and policies such as the new role of the European Parliament, the Council of PMs, the Economic and Monetary Union, all community policies etc. The two other pillars are not based on supranational competences as the first one, but in cooperation among governments. The second pillar is the Common Foreign and Security Policy (CFSP) – inspired by the EU's inability to handle the Yugoslav war – and the third refers to police and juridical cooperation.

From the beginning, the Treaty was deeply preoccupied by excessive national debts, thus the limit of budget deficits to 3 per cent of GDP was set up, and the ECB was specifically barred from bailing out national and regional governments. The budget deficit limit introduced austerity policies in all countries a decade before the crisis, and yet, fiscal policies is the only way for a country to respond to a sharp economic decline. Since all countries will be borrowing in euros, not in national currencies, i.e. without the ability to print currency, if borrowing gets too high, international speculative markets

(*Continued*)

may impose higher interest rates that could damage national economies. These rigidities and other important reasons kept away the strong economies of the UK and Denmark, while the fragile SE economies have been accepted into the Euroland, despite strong opposition by the left in these countries and without referendums.

optimum currency area. It ignored the relevant literature and experience on the subject and instead prioritised the economy at the expense of political integration, coming up with five national "convergence criteria". These are: no more than 1.5 per cent inflation rate more than the average of the three member states best performing in low inflation; government deficit must not exceed 3 per cent; the ratio of the annual government deficit to GDP must not exceed 60 per cent; applicant countries should have joined the exchange-rate mechanism for two consecutive years with no devaluations of national currencies before final entry; and long-term interest rates must not be more than 2 per cent higher than in the three lowest inflation member states.

In Europe, apart from the influence of British-style Thatcherism, Germany applied *ordoliberalism*, a German version of neoliberalism, developed since the late 1930s by economists and legal scholars at Freiburg University. Ordoliberalism became the dominant mantra of law and order with the following features: a strong central state intervention to secure competition, an independent central bank committed to monetary stability and low inflation a balancing tax revenue against government expenditure, privatisation of public services and public assets and deregulation of the labour market with some minimum wage (Aziz, 2015).[4]

In Eastern Europe, ordoliberalism has been painfully experienced via the bloody re-establishment of capitalism and accumulation of capital by extensively applied dispossession (Harvey, 2003). Extensive privatisations, land grabbing and abolition of any social right became the rule. In former Eastern Germany, all former state companies and properties have been transferred to a special agency, Treuhand, used as a model 20 years later in Greece. Whole operating factories were sold for a few Deutsch marks, or dollars, while massive unemployment and impoverishment devastated whole cities and the countryside. At the EU level, ordoliberalism became dominant among EU institutions, the founding principle for establishing the euro and later, when the crisis began, as the major tool to discipline the "wasteful and irresponsible" countries in the South via extreme austerity.

Since 1992, all EU countries have been forced to implement the Maastricht Treaty, which entailed abandonment of redistribution policies. In SE, governments were obliged to begin working towards the flexibilisation of employment relations (on top of existing informal flexibilities described in the previous section) and deregulation of public welfare. In addition, Greece and Portugal began to deregulate credit in the second half of the 1980s under the prospect of the single financial market. Spain and Italy had already initiated similar deregulations from the 1970s, while financial interventionism and financial neoliberalisation in the four countries was a state-driven pattern of reform in which central banks played a prominent role. The major change in Southern economic policies concerns the imposed deflationary regime by the ECB following ordoliberal criteria (Gambarotto and Solari, 2015). From growth driven by expanding internal demand, Southern economies were forced to apply a German style export-driven growth policy for which their productive basis – with few notable regional exceptions – was not prepared.

In Spain, Gonzales and the PSOE stayed in power until 1996, but they were open to pro-market development, implementing a version of Iberian-style neoliberalism with extensive cuts to social welfare, privatisations and tax cuts for the rich together with labour market deregulation. These measures caused strikes, which paralyzed the nation, forcing Gonzales to offer several concessions to end them. In 1992 the summer Olympics were held in Barcelona, and the World Fair was held in Seville, both symbolising the beginning of the Spanish real estate and financial boom. In 1996, the Aznar right-wing government came to power, and its first policy was to freeze civil servant wages and to introduce a strict budget. Aznar continued aggressive privatisations in telecommunications and energy, plus liberalisation of planning controls. Regional governments took this opportunity to give permits for extensive housing and real estate projects with immediate disastrous environmental effects and long-term catastrophic economic and social ones.

Italian politics in the 1990s started with institutional paralysis due to extensive corruption and organised crime's significant influence, collectively called "Tangetopoli". Major political parties lost the confidence of voters, and new coalitions emerged. The 1994 elections swept media millionaire Silvio Berlusconi into office as Prime Minister, but only for nine months. A series of short-life centre-left coalitions from 1996 to 2001 came to power,[5] all committed to neoliberalism, with strict budgets, employment flexibility and the abolition of minimum wages for some private and public sectors. Italy undertook a far-reaching privatisation programme in industries, utilities, telecommunication and banking, representing more than 125 billion euros in assets between 1992 and 2005 (Salento, 2014). In Greece, PASOK had been re-elected as governing party in 1993 and remained in power until the end of the decade. Prime Minister Costas Simitis applied neoliberal reforms with extensive privatisations and labour flexibility. The major task of Simitis' government was to prepare the Greek economy for entering the Eurozone. Instead of a rigorous plan, the administration engaged in mega projects for the 2004 Olympic Games with private capital in real estate speculations. This effort wasted valuable resources and time and remains until today the largest lost opportunity cost in a crucial decade.

Until the mid-1990s, the party in power in Portugal was Cavaco Silva's Social Democrats and from 1995 until 2002 the Socialist Party led by Oliveira Guterres. Both parties claim 80 per cent of the electorate in a nation of 10 million, and both see privatisations, welfare deregulation and labour flexibility as the necessary steps towards economic modernisation. More than 45 major companies were sold, including telecommunications, banks, insurance, breweries, newspapers and merchant marine companies.

Political elites in SE have traditionally had close contacts with economic dynasties, and individuals often circulate between business and politics, the classic case being Silvio Berlusconi in Italy. In smaller economies, such as Portugal and Greece, a limited number of rich families have remained at the core of their economies for the last 30 years of the 20th century and the early 2000s. In Portugal, they control particular sectors such as banking, media, telecommunication, energy and real estate (Marques, 2015); and in Greece sectors such as shipping, banking, oil and gas, media and real estate (Roth, 2013). These sectors and family clans were the main beneficiaries of government policies during the boom years before the crisis, enjoying low taxation and special contracts with the state, while transferring all profits to foreign banks and tax havens. Due to the small size of national markets, these economic clans used the state to pursue their targets, hence the key role of "friendly" political elites and political parties. In Greek, there is a particular term to describe these relationships, «διαπλοκή» (diaploki), meaning interweaving. Of course, the election of the real estate billionaire Donald Trump as president of the USA makes the SE cases looking like "peanuts" in the "diaploki" game.

Slow growth and problems in intermediate regions

Within this neoliberal political and macro-economic framework, the particular SE development pattern of SMEs, in combination with the performance of large firms, slowed down substantially, despite high rates of annual GDP growth, until the late 1980s. Subsequently, in 1996 it was 2.9 per cent in Greece, 1.3 per cent in Italy, 2.7 per cent in Spain and only Portugal showed as much as 3.5 per cent annual growth (OECD, 2003). During the 1990s, the gradual regional specialisation in consumer goods, produced with middle-range technology and relatively higher wages, plus regional specialisation in holiday destinations, showed its first limitations, and trade deficits started to predominate. Italy's total trade balance was positive until the late 1990s, when for the first time it reversed. In Greece, until the 1980s, imports/exports were more or less balanced, but from the early 1990s the trade balance became negative, and one can identify similar tendencies in Portugal and Spain also (Charnock et al., 2014).

Despite trade problems and the fact that uneven geographical development remained a key feature, a clear tendency for some southern regions to converge in terms of GDP, with EU averages, was visible. Convergence between regions in the EU-15 (measured by coefficient of variation) was strong up to the mid-1990s, but the process since then has slowed down, and since 2000, inequalities have been growing again, reaching 1987 levels in 2007 (Eurostat, 2007). The Commission identifies two speeds: convergence for regions with GDP higher than 75 per cent of EU-15 average and divergence for regions with GDP less than 75 per cent, the majority of which are in the South. Unevenness across the EU increased dramatically after the entrance of the former "socialist" countries of Eastern Europe. In 2002, the 10 per cent of the EU-27 population living in the most prosperous regions of North-Central Europe accounted for 19 per cent of total GDP for EU-27, compared to only 1.5 per cent for the 10 per cent of the population living in the least wealthy regions in the East and in the South (Eurostat, 2007). In adjusted prices, the ratio between the top and bottom 10 per cent of population in terms of GDP is 5:1 and in real prices 12.5:1. In 2008, 43 per cent of EU GDP was produced in only 14 per cent of its territory, within the geographical area defined by London, Hamburg, München, Milano and Paris, in which one-third of the union's population live and work.

These figures would be significantly more uneven without EU assistance programmes (Leonardi, 2006), particularly the structural funds and the application of the CAP. Following Todl (2000), structural funds for the period from 1989–1993 contributed to 2.71 per cent annual GDP growth in Greece, 3.39 per cent in Portugal, 0.71 per cent in Spain and 0.77 per cent in the Mezzogiorno. For the period from 1994–1999, the contribution to annual GDP growth was respectively 2.82 per cent in Greece, 3.26 per cent in Portugal, 1.30 per cent in Spain and 1.14 per cent in Mezzogiorno. Portugal and Greece seemed to benefit more, and this was also evident in their overall performance during these periods. At the regional level, structural funds have been particularly effective (comparing GDP/head figures for the period from 1980–1994) in Canarias, Extremadura, Aragon and Navarra in Spain; Algarve, Norte and Alentejo in Portugal; and the Northern Aegean islands, Crete, the Ionian Islands, Eastern Macedonia and Thrace in Greece. These are mainly agricultural and insular regions with low initial GDP/head figures and remoteness problems and have been growing faster than the more developed areas.

After the mid-1990s, however, the regional problem in SE shifted to *intermediate regions* identified from the 1970s up to the mid-1990s as dynamic and innovative, as a "third way" of regional growth beyond metropolitan areas, state-assisted industrial growth poles and backwards agricultural areas

(Paci, 1972; Garofoli, 1983; Hadjimichalis and Papamichos, 1990; Amin, 2003). As described in the previous section, these were successful regions in Third Italy, used as models for local/regional development. Other successful regions included Northern Greece, Valencia, Basque Country, Murcia and Northern Portugal, among others, which for the first time showed signs of slow-down. Thus, from the mid-1990s onwards, and in the context of the wider neoliberalisation and financialisation with recession as an outcome, several major negative changes gradually took place in SE economies and particularly in their competitive intermediate regions. The following were especially important.

On the domestic front, the gradual neoliberalisation of state and regional institutions changed the supportive framework for SMEs dramatically, with extensive privatisation of public services on various spatial scales. Between 1990 and 2000, privatisations in the EU, as a percentage of GDP, were led by Hungary with 2.7 per cent, followed by Portugal with 2.3 per cent, Czech Republic 1.3 per cent, Greece 1.2 per cent, Poland 1 per cent, Ireland 0.9 per cent, Italy 0.8 per cent and Spain 0.7 per cent (*Financial Times*, 28 April 2001). In effect, state and regional supportive institutions gradually withdrew their support by asking SMEs to pay high fees for previously inexpensive contributions, e.g. for technical assistance. Furthermore, EU regulations prohibited national, regional and local incentives to all types of firms, being against competition at the European scale.

Another equally important withdrawal of domestic support for the SE development pattern was in the social sphere and relates to younger family members and particularly women. The family remains the traditional cornerstone of Southern SMEs' success, and its continuation is essential for the survival of small firms. But across SE, in all sectors, a gradual loss of interest in reproducing themselves as skilled and unskilled workers in small family firms became widespread among younger generations, particularly among women (Vaiou, 1997). Many surveys have found the same lack of interest among the sons and daughters of small entrepreneurs, who were unable to convince their epigones to continue the small family business (Mingione, 1998b, 2009; Nesi, 2010). The younger generation prefers higher education or working in services, avoiding the long hours of hard and unstable work with low pay that is typical in most small firms. Furthermore, changes in consumption patterns and the increase in divorce rates and short-term social relationships changed employment relations as well, as younger people were forced into precarious work to make a living (Miguélez and Recio, 2010).

To this, I must add severe demographic problems across SE, which in some regions is synonymous with total depopulation and growing ageing. Without a new influx of young people, the interior of Greece, Portugal, Italy and Spain will be practically empty by 2030, due to depopulation. These four countries faced three serious problems from the 1980s onwards: low birth rates (below 1.5); high levels of population over the age of 65 (in the 1990s Italy had 22 per cent, Greece 20 per cent, Spain 19.6 per cent and Portugal 13.1 per cent); and higher proportions of retired people relative to those working, due to unemployment and undeclared employment. This dependency rate will grow, on average for the four countries, from 50 per cent to 57 per cent by 2021 and in combination with unemployment could turn to a "demographic death trap" with "childless and jobless" economies. In the 1990s, unemployment was around 7 per cent–12 per cent in Greece and Italy, 14 per cent–22 per cent in Spain and only Portugal had as low as 5.5 per cent–7 per cent. The ageing population gave rise to new care needs that families were unable to sustain. The combined effect of these problems was higher demand for public social services, particularly pensions, health care and unemployment benefits and the creation of a new labour market demand for care workers that was filled with informal immigrants' labour (Cánovas and Riquelem Rerea, 2007). All the above phenomena fuelled governments' fiscal problems.

Turning to productive restructuring, distinct from deflationary policies, in the 1990s, large and small firms in SE faced severe problems in pricing their products. Each specialised activity in which the productive process was subdivided had a local/regional price system, related to sectoral and product specialisation, to local productivity and capital-labour relations and to other local economic, social and cultural characteristics. This local/regional price system, however, was open to competition with other price systems in Europe and globally. Labour cost remained a key component in the price system due first to the labour intensity that characterises Southern productive processes, second to the social organisation of production, and third to the relative immobility of labour. Particularly for the majority of small firms, which are price-takers, product-followers and react to innovation by others, their local/regional embeddedness turned to a trap when other producers from other places started producing the same products with substantially lower labour costs. In the late 1990s to the early 2000s, the average cost per hour for an Italian worker was 17.823$, while in neighbouring Slovenia it was 6.7$, in Romania 4.3$ and in China was less than 1$ per hour (*La Repubblica,* 12 September 2004).

Another important endogenous missing organisational device in most Southern firms was crisis management. The majority of small firms tended to remain confident in old-fashioned unwritten agreements and informal face-to-face communication, which were functional during the golden period of the 1970s and 1980s but highly problematic when the changes analysed above occurred. Most firms in IDs, especially in Italy and Spain, were lacking in institutional devices able to promote a rapid selection of new leaders and crisis managers (ARMAL, 2003; Ybarra et al., 2004) despite the presence of many supportive local institutions. As a result, during the prolonged period of slack demand and due to higher pressures from international competitors, internal competition became primarily price competition and hit the weakest categories, i.e. small independent firms, subcontractors and homeworkers.

On the exogenous supranational and international front, the world recession of the late 1980s to the mid 1990s initiated a substantial reduction in demand for consumer products produced in SE intermediate regions (e.g. design and fashion products), particularly from the USA and Germany, a situation that deteriorated dramatically after 9 November 2003. At the same time, due to geopolitical changes after 1989, new markets and new competitors in Eastern Europe, Northern Africa, Turkey, India, Vietnam and above all China entered the scene for low-medium quality garments, textiles, ceramics, toys, footwear and furniture (Smith et al., 2002; Labrianidis, 2008; Dunford et al., 2013). The economic opening of eastern countries was positive for some Southern firms, which found new markets and new territories for foreign direct investments, but for the majority, it was a negative development resulting in massive closures, or in de-localisation. A major shift in the European division of labour gradually took place, because southern regions lost their preferential economic relations vis-à-vis central-northern regions, as those relations shifted to Eastern Europe. This has been evident in the case of Germany. At the beginning of the 1990s, German imports from EU countries and from SE in particular were on average 12 per cent of total imports, while from Eastern Europe they were on average 8 per cent (Simonazzi et al., 2013). At the end of the first decade of the 2000s, it was 9 per cent from SE and 13 per cent from Eastern Europe. Imports from Eastern Europe were coming from de-localised German firms, which succeeded in sustaining competitive productivity and production costs while contributing to the sharp fall in Germany's relative unit labour costs (Simonazzi et al., 2013).

These developments at the global, European and national levels, in addition to closures, pushed many micro-, small and medium firms in

Box 2.4 Mergers and acquisitions in the Italian textile-clothing sector and in the Greek food sector

In Italy, older cases of acquisitions that formed powerful large integrated firms, like Benetton, Gucci, Stefanel, SASIB and others, have been well discussed. But after 1998, new ones, like Diesel (clothing) and Marzotto (textiles) in Veneto and Prada (footwear, leather, clothing and other fashion items) in Piedmont and the Marche, are good examples of how some innovative firms with brand names in former local networks bought their partners and formed large, integrated companies with subcontractors. Between 1980 and 1995, in Veneto alone, 2000 acquisitions took place, reducing the total number of firms by 13 per cent and increasing the ratio of employees/firm from 5.6 to 25.7 (Fondazione Nordest, 2003). In Greece, the food sector was the target for many international firms during the period from 1987–1997. Nestle-Suchard from Switzerland and Sara Lee from the Netherlands acquired many local chocolate industries, such as Loumidis and Ion. In the pasta sector, the Italian firm Barilla bought the Greek giant Misco, and in the meat sector, the German firm Moskel bought Marox. Milk and yogurt, however, remained in Greek hands and formed a powerful duo: Fage and Delta, which also became international brands (Skordili, 1999). These acquisitions substantially reduced, by 12.5 per cent, the total number of firms in the sector and also the number of workers through modernisation. All acquired firms kept their brand names and were used by foreign firms as entrance points to the Balkan and Turkish markets, where Greek firms have long-established networks (Skordili, 1999).

SE towards three interrelated restructuring processes: mergers, acquisitions and the formation of large, vertically integrated firms and groups of firms, mainly in the manufacturing, tourist and construction sectors; the de-localisation of part of production, or of whole firms, including many small and medium firms, to regions and countries with low labour costs, in all manufacturing subsectors; extensive replacement of local craft-workers by non-EU migrants to compensate for increasing local labour cost; and/or the lack of skilled labour in all sectors, including agriculture.

The first restructuring process was already visible from the late 1980s (see Bianchi and Bellini, 1991; Harrison, 1994) and was widespread across all sectors and regions. The prospect of the single market and the gradual disappearance of trade barriers strengthened mergers and acquisitions. Banks, insurance, chemicals, telecommunication and car manufacturing led the way, followed by the textile-clothing and the food sector. The other two restructuring processes, de-localisation and the use of immigrant labour, are part of the major restructuring of the European and global spatial division of labour and require further scrutiny.

De-localisation and changes in the European and global spatial division of labour

> ....It is not only the decline of GDP during the last 8 years which worries me, but above all what alarms me is the decline of Italy's share in international trade, which shows a reduction by 25% from 1995 to 2003.
>
> Carlo Azeglio Ciampi, President of Italian Republic, speech on 1st May 2004, (*L'Unitá,* 3 May 2004, author's translation)

"Italy fights to remain home of luxury fabrics and fashion products"
Headline of an article in the *New York Times,* 15 March 2004

Although the above quotes concern Italy, they are representative of the problems visible in the 1990s and early 2000s in other parts of SE as well (see also Harrison 1994; Hadjimichalis, 2006; Pickles and Smith, 2015). From the beginning of the 1990s, Southern firms in all sectors (particularly in fashion products, (TCFFT) Textiles, Clothing, Footwear, other leather products, Furniture and Toys) have started, parallel to the export of finished products, a process of de-localisation of parts or all production away from southern regions to places and countries with lower labour cost. De-localisation took two forms: direct use of foreign subcontractors in remote destinations, especially for less complex phases of production (India, Vietnam, China) mainly by Italian and Spanish firms and to a lesser degree by Greek firms; re-location of machines and some native technical supervisors to closer destinations (Eastern Europe, the Balkans, North Africa) by firms from all four SE countries. Relocating tasks in new destinations was more complicated, using the cheaper but equally skilled labour force in the new location.

Capital always searches for cheaper production locations, where social and environmental conditions are favourable and labour is docile and productive. Some capital fractions are more mobile across space and national boundaries, while others remain fixed. The interplay between motion and fixity is one of the main pillars of uneven geographical development (Harvey, 1982). However, capital cannot move freely across space by itself; it requires regulations and the strong involvement of the state to secure the whole process. From the late 1990s, the EU promoted two parallel selective strategies to help the mobility of capital: an uneven enlargement of the EU itself towards Eastern Europe and a deepening of economic and geopolitical integration with neighbouring territories in the Mediterranean that were not likely to become Member States. Both strategies have done nothing, or very little, to change the uneven economic development between the EU, Eastern Europe and the Mediterranean coast of Africa, as they were designed to promote EU capital interests. The end result has been a differential inclusion/exclusion from EU economic space (Smith, 2015).

The EU's trade policy, particularly the (OPT) Outward Processing Trade agreement, was a major driver behind the thriving intra-European apparel trade, after the capitalist transformation of Eastern Europe (Plank and Staritz, 2015). It established a particular spatial division of labour where former communist countries provided labour-intensive production options, allowing EU-based firms to temporarily export inputs for processing and re-import products under preferential conditions. These agreements favoured the apparel and footwear industries (in France, Italy, Spain, Portugal and Greece) and mechanical products (in Germany and Austria) (Cutrini, 2011). This led to a massive class, gender, age and ethnic re-composition of

Table 2.2 Imports of semi-finished and final products (sector TAC) in Italy according to areas of origin (in millions euro, %, 1991–2001)

	1991		*2001*		*1991–2001*	
	From high labour cost regions	*From low labour cost regions*	*From high labour cost regions*	*From low labour cost regions*	*From high labour cost regions (%)*	*From low labour cost regions (%)*
Clothing	1.328	1.076	2.179	5.580	64.0	418.4
Textiles	2.476	1.028	3.122	2.782	26.1	170.8
Footwear	282	392	619	2.438	119.1	521.3
Total	4.087	2.496	5.919	10.799	44.9	332.6

Source: Fondazione Nord Est (2003).

the EU and Mediterranean labour markets, where national and local barriers to entry were removed for the benefit of particular capital fractions.

Eastern Europe's rising share in the EU-15 market from the 1990s was driven by several factors, apart from the OPT trade agreement, including a relatively cheap but well-trained and disciplined labour force, proximity to Central Europe, time and flexibility in production and the existence of large, vacant industrial facilities. In the 1990s and early 2000s, strong increases in TCFFT trade with Romania, the Czech Republic, Poland and Bulgaria in Eastern Europe and Tunisia and Morocco in North Africa defined the emergence of a new Euro-Eastern-Mediterranean integrated zone, shifting production to these new locations. Later, China, Turkey, Pakistan and Vietnam extended the outsourcing zone to the Far East, negatively affecting exports and employment in Eastern Europe.

In 2004, the EU as a whole lost 160,000 jobs in the TCFFT sectors, and in 2005, another 164,000 jobs were lost. Capital in these sectors started to worry, and its EU association/lobby (EURATEX) pushed for safeguarding measures that came into force in June 2005 with a special agreement with China to restrict some of its exports. At the turn of the millennium, and due to changes in international trade regulations, an extraordinary capital restructuring took place in TCFFT sectors with shifts from Southern to Eastern Europe and the Mediterranean and then to the Far East. It shows, once more, that capitalist success in particular locations is only temporary, as capital moves freely over space, while labour, being less mobile, faces the negative externalities in situ.

So, the catchphrase *Tutti all' estero* (All abroad), which was popular among Italian entrepreneurs in the 1990s, echoed the aspirations and practices of many entrepreneurs in SE as well, although not all were able to make this a reality, or to take the opportunities to upgrade their activities. Most firms do not de-localise to get access to new markets, but are simply engaged in "a race to the bottom" to face the competition from low-wage countries. In the case of Italy, which led the process, a good indication of the extent of de-localisation is data for imports of semi-finished and final products from high- and low-cost regions. In Table 2.2, the data is divided into two periods: 1991 and 2001. The dramatic increase in import figures from low-cost areas in 2001 is remarkable in all three sectors, as well as the gradual penetration of the domestic market by foreign footwear, mainly medium-quality shoes.

In the beginning of the 1990s, Hungary and Slovenia were the main destinations for the TAC sector from Italy and particularly from Veneto, as Bulgaria was from Northern Greece and Morocco from Southern Spain. They were the closest countries with adequate infrastructures, and a qualified, well-trained, but low-cost, labour force (Crestanello, 1999). Later, Italian firms went to even lower cost countries like Romania, Poland and Belarus. In Asia, they went first to regions specialising in textiles and clothing, like Hong Kong and Taiwan, but later to China, Thailand and Vietnam, searching always for cheaper and more productive labour. From the 5,643 Italian firms having investments abroad, 2,000 are real de-localised firms, i.e. they have transferred all or a substantial part of production to lower-cost areas. One small entrepreneur describes the process:

> ...an Italian garment producer may today import textiles from Turkey produced there under Italian quotas, send it for cutting and stitching to Romania, realize the final phases in Italy and export the final product to US under an Italian brand name, different from their own
>
> (interview by author, Vicenza, March 1999)

Spain, from 1990 to 2004, registered 130 de-localisation cases of firms with more than 500 workers in all sectors. In this period, offshoring was more important in some sectors with high-medium technology such as electric and electronic machinery, motor vehicles and in labour intensive sectors such as textiles, clothing, footwear and the wood and cork industry. This was clear in the index of dependence on the import of intermediate goods in the manufacturing sector, which increased from 7.2 per cent in 1990 to 11 per cent in 2005, with an increase of over 50 per cent during the entire period (González-Diaz and Gandoy, 2005). Particular capital fractions such as EUROMETAUX, the lobby defending the interests of non-ferrous metals capital in Brussels, asked the EU for compensation to save 33,500 jobs in Spain from potential de-localisation.

In the Spanish clothing sector, there were 300,000 jobs in the 1990s, while in the mid-2000s there were scarcely more than 135,000 as a result of companies relocating their output to Asian countries, particularly to China. However, after two decades of the de-localisation boom, some production that moved abroad returned home, according to the Federation of Spanish Clothing Companies. Spanish companies are now looking to manufacture closer to home in Portugal and Morocco. Quoting INE (Instituto Nacional de Estatística), in 2003 alone, 94,800 jobs were lost in the industrial sector. Textiles, clothing, footwear, electronics and automobiles – which in Spain accounts for 8.5 per cent of GDP, 13 per cent of employment and 30 per cent of exports – are the most affected sectors. Moreover, well-known firms such as Samsung, Phillips, Panasonic, Nissan, Volkswagen, Zara and Mango have decided to close down their Spanish plants or relocate some of their activities abroad (Charnock et.al., 2014).

In Greece, the textile-clothing industrial cluster in Northern Greece was gradually shifting more to the north, crossing the borders to include parts of Southern Bulgaria, Southern Albania and FYROM (Former Yugoslav Republic of Macedonia) (Labrianidis, 2008). In the 1970s and 1980s, Northern Greece textile and clothing firms enjoyed a preferential trading relationship with Germany, the Netherlands, and to a lesser degree with the UK and France, as subcontractors of medium quality products. This

Box 2.5 Moving frontiers

In important research, Bertoncin, Marini and Pase (2009), *Frontiere mobili*, describe processes of de-localisation in both places of origin and destination, from the IDs of Veneto to Timisoara in Romania, to Šamorín in Slovakia and to export platforms in Tunisia. From the beginning of the 1990s, the old ID of Montebelluna in Treviso, which is specialised in sport footwear, attracted giant multinationals such as Adidas, Nike, Salomon and Rossignol, contributing to in situ internationalisation. At the same time, the most advanced and large Italian firms of the Sport System of Montebelluna started de-localising part of their production abroad, mainly to Timisoara, Romania and Tunisia. Until 2000, 16 per cent of the firms had been de-localised, while by 2006 the figure had increased to 28.9 per cent. In terms of size, by 2006 100 per cent of large firms had been de-localised, 78.6 per cent of medium firms, 38.4 per cent of small firms and 16.4 per cent of micro-firms. Gradually, de-localisation went farther away, from Eastern Europe to Tunisia and later to the Far East, to China and India (Buzzati and Passquato, 2009).

Timisoara in Romania attracted so many firms from Veneto and Treviso in particular, that its nickname among Italian entrepreneurs is "Trevisoara". Timisoara's attractiveness includes the high qualifications of the Romanian labour force for labour-intensive tasks in shoe production, the existence of old, large former socialist shoe factories that were used during the first period of de-localisation (1990–1995), the absence of environmental controls, lower labour costs and proximity to an international airport. Later, Italian firms came with the "whole factory in one TIR" and many Italian technicians and supervisors to control production quality (Scroccaro and Sivieri, 2009).

can largely explain the important role of clothing and textiles in Greek exports during the same period. Low labour costs, efficiency and delivery time explains this trade relationship. From the 1990s, due to competition with eastern countries, Greek companies started using third party subcontracting, resulting in a gradual shift of more complex production tasks to the Balkan neighbours, and initiating what Kalogeresis and Labrianidis (2008) call "triangular manufacturing". Proximity, cultural affinity and "knowing how to do business in the Balkans", played crucial roles in fostering de-localisation from Greece towards Southern Bulgaria, Albania and FYROM. These shifts have had negative regional domestic effects: a 23.8 per cent decrease in employment and a 43.8 per cent decrease in firms in the textile-clothing sector during the period from the mid-1990s to 2004. After 2004, the decrease was more dramatic: 70.3 per cent in employment (mainly middle-aged women) and 83.2 per cent in the number of firms. From a total of 81,000 jobs in the late 1980s, the sector went down to 15,000 jobs in 2004.

If the eastwards enlargement was less beneficial for SE firms, due to their productive specialisation and technological deficiencies, this was not true for German firms. From the mid-1990s, the pattern of German de-localisation to Eastern Europe differed from SE in two important aspects. First, it concerned manufacturing activities of intermediate products in mechanical sectors (e.g. car parts), which helped to create jobs in Eastern Europe while contributing to the sharp fall in Germany's relative unit labour cost. Second, German de-localisation kept the final stages of production at home, unlike most de-localisation from SE. The German export success was the result of major internal labour relations reforms in the 1990s and of outsourcing practices, which, via lower wages and prices, created a competitive supply chain for German industries. Cheaper imported intermediate goods from Eastern Europe supported Germany's export-led growth in technologically advanced products, contributing further to the marginalisation of southern regions. As Simonazzi et al. (2013) argue:

> ...the increasing integration of Central and Eastern European economies within the supply chain of German industry speeded up their process of diversification-cum-specialization. The eastward integration of German industry, combined with the persistent containment of internal demand in the major economies of the euro area, has gone hand in hand with the impoverishment of the productive matrix of those southern regions less connected with Germany
>
> (p. 664)

Economic migrants in SE: from *gastarbeiten* to migration of multi-ethnic clandestine labour

Gastarbeiten is the German word for guest workers, used by Central Europeans to describe the massive migration from SE countries and Turkey mainly to Western Germany but also to France, Belgium, Austria and the Netherlands in the 1960s and 1970s to work in their factories and mines, following bilateral agreements. At places of destination, at work and in everyday life, Southern immigrant workers were confronted with racism and extreme exploitation, while their labour made possible the "miracle" of fast growth after the devastation from the Second World War in Central Europe. Between July 1959 and September 1961, 549,000 foreign workers worked in the Federal Republic of Germany, but not dark-skinned Portuguese.

During the 1980s and 1990s, dramatic changes took place in SE regional labour markets. In two decades, southern regions received thousands of foreign migrants, creating profound economic, social and cultural transformations. From places of emigration, they became places of immigration, without having any experience in accommodating multi-ethnic populations. Although reliable estimations are extremely difficult, data on migrant populations in 1998 from OECD -International Migration Outlook (2013), supplemented by EUROSTAT data, shows that in the 1990s, Greece received 700–900 thousand, 9 per cent to 10 per cent of its population; Italy 1.5 million, 2.6 per cent of its population, Spain 900 thousand–1.2 million, 2.3 per cent of its population and Portugal less than 300 thousand, 2.8 per cent of its population. In 2004, after detailed censuses in all EU countries, Greece had the highest ratio of migrants to population, at 9 per cent, followed by Spain at 6.4 per cent, Germany at 5.8 per cent, Italy at 4.3 per cent and Portugal at 3.1 per cent, while all other countries had lower rates.

Changes in the European and global division of labour are responsible for this massive mobility of people searching for a better future in Europe. In the countries of origin, overpopulation, unemployment, wars and civil wars, plus environmental problems, explain in part their decision to risk their lives crossing the Mediterranean and traversing mountains. The choice of SE (as entry to continue to Central-Northern Europe or as final destination) is explained by several factors. First, the particular endogenous development pattern, described previously, creates a demand for cheap and flexible labour for low-skilled tasks. Second, the existence of a vast informal sector provides a window of opportunity for clandestine work. Third, the relative lack of migration experience in the four countries resulted in less rigid migration laws and in poorly organised institutions. Fourth, demographic problems, family transformations and declining social protection generated a particular labour market for domestic services, for which female immigrants provided solutions. Finally, easier access from the sea for all four countries and from the mountains – in the case of Greece – provided many difficult to control entry points (Venturini, 1988; Pugliese, 1995; Anthias and Lazaridis, 2000; King, 2000).

Due to complex bureaucracies and inefficient management, it is difficult to estimate – some argue whether it is even desirable – the number of legally present migrant workers, particularly in Greece but also in the other southern countries. Table 2.3 shows the distribution of migrant employment by

Table 2.3 Migrant employment by economic activity, 1998–1999 (in %)

	Greece	*Italy*	*Spain*	*Portugal*
Agriculture/fishing	3.5	6.0	9.0	3.3
Manufacturing/mining	19.3	29.0	11.6	17.4
Construction	26.6	9.4	8.8	18.6
Retail wholesale	19.0	17.7	26.1	24.3
Health social services	5.9	11.1	14.2	17.9
Home care	19.9	10.4	16.4	5.1
Public administration	0.8	3.0	1.3	1.8
Other services	5.0	13.4	12.5	10.6

Source: OECD (2001).

economic activity in 1998–1999 based on the work permits granted (see also Baldwin-Edwards and Arango, 1999).

Taking agriculture as an example, a common feature in Table 2.3 is that the figures underestimate the presence of thousands of migrants working without permits in many southern regions, such as in Almeria, Spain in greenhouses producing vegetables, in Ilia and Arta in Greece for strawberries and oranges and in Latina and Caserta in Italy for fruits and grapes. Eurispes (2001) estimates that in Italy, 38 per cent of all non-EU migrants were employed in agriculture, in Spain, 29 per cent and in Greece, 33 per cent, and many reports have documented slavery conditions in these regions (*International Herald Tribune*, 26.1.1012). Underestimates are also present for home care and other services that provide employment for women coming from Eastern Europe, Latin America and the Philippines. On the other hand, in manufacturing and mining, due to relatively effective regulations, higher figures can be seen.

All SE researchers, however, emphasise that these official figures do not represent reality (Pugliese, 2002; Vaiou et al., 2007). The structure of the labour market with its many "holes" for informal work provides benefits for employers, such as avoiding social security payments, paying less for migrant labour and doing little for health and safety (Baldwin-Edwards and Arango, 1999). In Italy, despite a strict law on migration introduced by the Berlusconi government in 2012 (after pressure from the extreme right), boatloads of desperate people kept coming, and in the 1990s, 2,000–3,000 lost their lives in boat accidents. In Italy, as Lanziani (2003) noted, of particular interest is the distribution of the migrant population towards Italian IDs with small, historical urban centres, where they find both housing and employment:

> ...where industrial districts faced a crisis, migrants became the solution for many SMEs especially where restructuring was synonymous with cheap and semi-skilled labour...The influx of immigrants helped

> particularly two categories of small firms (a) firms with productive cycles which were either expensive or risky to relocate to other zones, (b) subcontractor firms totally dependent on cheap labour….In any case the arrival of migrants initiated a process of economic and social metamorphosis in small urban centres with particular effects at the micro-level of everyday life…
>
> (Lanzani, 2003: 15–16, author's translation)

This metamorphosis described by Lanziani is also visible in most southern cities today, with migrants forming the majority in certain labour intensive sectors, such as street trade and cleaning and care jobs. This has been a real cultural change (some speak of a shock) for countries without experience

Box 2.6 Factories in the fields: racism in El Ejido and Nea Manolada

El Ejido is a small town of 50,000 Spaniards and around 30,000 migrants, some with work permits others without, mainly from Morocco and Algeria. It is the centre of the rich agricultural region of Almeria, known as "California Andaluza" (Andalusian California), the largest concentration in Europe of intensive vegetable production in greenhouses. Low-paid, hard and unhealthy work in greenhouses predominates, such that labour fraud, overexploitation and poor housing have been features of El Ejido. Production for export is organised with many similarities to the ID model, adapted to agro-industrial needs (Martínez Veiga, 2001). In early February 2000, the small town was the scene of the most violent outbreak of racism in the recent history of Spain. During two days and nights, local Spaniards attacked houses, shops and mosques – looting and burning – after a mentally disturbed migrant was accused of murdering a Spanish woman. The response by migrants was to call an indefinite strike to demand compensation for damage and the legalisation of workers without permits. After a week, the central and regional governments agreed to meet these demands, while the local right-wing municipality showed indifference.

In Nea Manolada, Southeast Peloponnese, 90 per cent of Greek strawberries are produced – 85 per cent of them are for export, making Greece the world's 24th largest producer in 2011. In the 1990s, migrant workers began to arrive in the area to work in the greenhouses, and by 2010, there were 3,500–4,000, mainly Bulgarians and large numbers of Pakistanis and Bangladeshis. "...The workers live in the fields, in sheds made of tin, paper and nylon under fairly inhumane conditions" (Gialis and Herod, 2014: 142). Often migrants are victims of racist behaviour when they visit the town, so the landowners developed informal "company shops" profiting from migrants' restricted consumption. Immigrants received around 20–23 euros for 12–15 hours work in 2011, and local farmers delayed the payment until

(*Continued*)

the end of the harvest. These conditions led to the first immigrant strike in Greece, in 18 April 2008, when hundreds of workers gathered in the town's small square protesting against low pay, poor working and living conditions and several months' delay in payment (*Eleftherotypia*, 19 April 2008, in Greek). Thugs hired by farmers attacked the workers during the night, leaving several of them seriously injured. This attack generated a solidarity movement across the country. Several activists arrived in Nea Manolada, and additional mobilisations occurred in Athens. After 20 days of strikes and negotiations, workers gained a 20 per cent salary increase. Although this was a major victory, the Nea Manolada experience generated aggressive behaviour by employers across the country.

in accommodating foreign cultures. The migrants have also revitalised neighbourhoods with old building stock by renovating abandoned houses and apartments, usually of lower standard. In de-populated agricultural regions, particularly in Greece and Italy, the arrival of migrants to work in agriculture and construction was a vital demographic contribution, giving life to small villages, keeping primary schools open and sustaining some local consumption.

Migrants are employed in positions where natives don't like to work – in dirty, difficult and dangerous jobs (the so-called "3-D jobs") as well as those with poor social status (Reyneri, 2001) – and they face severe discrimination, racism and exploitation (see Box 2.6) and fuelled the rise of racist political parties such as Lega Nord in Italy, LA.OS in Greece and since 2000, Golden Dawn. They are caught in the contradictory attitude of being "wanted but not welcome". The undocumented status is considered as non-existent in respect to their daily relations with authorities and social services. In Italy, the anti-migrant Northern League campaigns against migrants, with slogans such as "Yes to Polenta No to Couscous", while at the same time local industrial associations lobby for more migrants to work cheaply in their small firms.

The presence of foreign migrants was not without protest from local unions. In Modena, Vicenza, Prato and Bergamo, local syndicates in clothing have strongly protested against the presence of Chinese workers who work with Third World salaries in small firms (Nesi, 2010). Unions in Carpi estimated that by the late 1990s, in local textile and clothing firms, there were 1,600 Chinese workers (one-third of the employment in the two sectors), although only 850 were registered (*Il Sole 24 Ore,* 2002). According to OECD (2013), the net fiscal position of migrants was less favourable than that of the native-born population, but their net fiscal impact is extremely difficult to estimate, as calculations are based on registered employment and consequently on their contribution in paying taxes and social security. Using different assumptions and formulas, OECD (2013) concluded that the net contribution of migrant households (taxes plus social security payments minus social transfers they receive) for the period from 2006–2008 was positive in the four SE countries and above the OECD average. The positive balance sharply declined after the 2010 crisis. These estimations, however, concern only migrants with working permits and leave out those millions working in the informal economy. In this respect, the higher the informal economy, the larger the proportion of native-born and migrants who do not contribute to taxes and social security payments, while their labour and surplus value production contributes to GDP growth. When one considers that estimations for the informal economy are 24 per cent–31 per cent of GDP in Greece, 21 per cent–28 per cent in Italy, 19 per cent–23 per cent in Spain and 17 per cent–21 per cent in Portugal, then the loss of revenue from undeclared work of both natives and immigrants is very high (Eurispes, 2001; Eurostat, 2006).

Another parameter is migrants' length of stay: many studies have shown that the more years they stay in a country, the more they are likely to obtain legal status and thus pay social contributions (OECD, 2013). Thus, countries with migration experience and more progressive legislation concerning "legalisation" may enjoy higher contributions from working migrants. This is the case in Italy, where, during 2014, migrants contributed roughly 13.3 billion euros to the Italian state in taxes and social security, while the government spent 11.9 billion on migration measures, leaving a surplus of 1.4 billion (*The Italian Insider*, 13 November 2015).

* * *

In summary, the particular/different development path in southern regions during the 1980s and 1990s is characterised by several factors, among which I note the following: the persistence and innovation of agrarian social structures combined with modernised tourism and higher rates of small and very small family business and informal work compared to North-Central Europe; the limited proletarianisation of the manufacturing working class, the high presence of migrant labour and the strong state interventionism followed by clientelism and patronage; the concentration of large industrial activities in only a few urban regions and the diffused patterns of flexible specialisation of modern small craft industries; and, finally, inflationary fiscal policies. In the 1990s, the increasing relevance of large companies from mergers/acquisitions and of foreign capital investing in privatised public utility companies, tourism, telecommunication, aviation and real estate, changed considerably the productive structure in SE.

After the Single Market, the Maastricht Treaty and the institutionalisation of neoliberalism, the structural weaknesses of SE capitalisms became more acute. On the one hand, the gradual hollowing-out of

European states by transferring major regulatory and institutional powers to EU bodies badly affected southern states addicted to strong interventionist policies. On the other, strong deflationary policies and the loss of various trade protections, imposed by EU regulations, destroyed the backbone of Southern economies, namely the networks of small-medium industrial enterprises increasing the role of large foreign firms now using SE as a profitable platform. These developments, plus the major restructuring of the European spatial division of labour, contributed to the slowing down of productive economic activity, reducing public revenues and raising state borrowing. The slowdown of GDP growth increased the ratio of debt-to-GDP, the Achilles heel of Southern economies. Thus, SE regions and states entered the Eurozone ill prepared and with different and weaker economic and institutional structures compared to Central European states. This soon resulted in deepening uneven development between and within countries, which became, in 2009–2010, a major crisis-driven process.

Notes

1 In the 1960s and 1970s, the creation of industrial growth poles with steel, shipyards, oil refineries, petrochemicals and other heavy industries became a dogma among regional development planners. In the same period, innovations in maritime technology and possibilities to transport the raw material necessary for steel from all over the world, plus using imported oil as energy, made costal locations more cost-effective than inland ones. In that period, growth poles found extensive application in *port-industrial complexes* across the Mediterranean. For more, see Rossi (2009) and Dunford and Yeung (2009).

2 The lira in Italy was devalued four times during the period from 1960–1992, the peseta four times in the period from 1976–1995, the escudo three times during the period from 1983–1995 and the drachma three times during the period from 1983–1991.

3 After the war, in Southeast Kosovo at Uresevic, the USA built the military Camp Bondsteel. It is a 1,000-acre unreal city for 8,000 people, created by Halliburton, and operates as an outpost controlling oil flows from the Caspian Sea and the rich mineral reserves of Kosovo valued at 300 billion dollars. A consortium of Evidity (a Canadian mining company) George Soros and Sahit Muja (ex-member of Central Committee of the Albanian Communist Party) guarantees the success of the project. See www.wsws.org/29.4.2002/PaulStuart.

4 For ordoliberals, the State is necessary to achieve the market ideal, but a particular kind of State. They have a limited faith in democracy and instead they strongly believe in independent, networked institutions that are unaccountable and operate parallel to the State. The trouble with ordoliberalism, as Aziz (2015) notes, is that when situations/basic parameters change, the rules of independent institutions stay the same, and if that means the problem is not solved, then so be it. This is how the EU and Eurozone operate.

5 The right-wing coalition, led by Forza Italia and dominated by Berlusconi, includes National Alliance, a crypto-Fascist party, the xenophobic Northern League and the Union of Christian and Center Democrats. The center-left coalition, called the Union, included the Democrats of the left, Greens, Christian Democrats, moderate communists, Rifondazione Comunista and the Daisy.

3 Uneven development II

Capitalist transformation and the building of the Eurozone

> ...The crisis of the Eurozone has its roots in geography and in the inability to take geography seriously. The failure to construct a financial architecture that could adequately work with the pre-existing economic uneven development between countries led to a further deepening of that economic inequality.
>
> Doreen Massey, 2012

When the unequally developed southern states and regions found themselves with the same hard currency in the 2000s, very few in SE and in the European Commission paid attention to their pre-existing highly unequal regional production systems and specialisations, to their structurally different regional labour markets or to their unequal accessibility to markets (economically, institutionally and spatially) vis-à-vis the 'core' of the Eurozone – this under the influence of the ordoliberal doctrine. Even fewer paid attention to the socio-spatial effects of putting an unevenly developed South into a macroeconomic and fiscal framework designed specifically for Northern-Central European economies and particularly for Germany. This constituted a kind of imposed straightjacket.

In this chapter, I challenge the dominant view that debt, public and private, is the sole cause of the crises in SE. Instead, I shed light on five important and highly interlinked developments that preceded the debt crisis and created the conditions for its emergence as a major problem. Public debt, or more accurately the ratio of public debt to GDP, is the outcome of the crisis that resulted from the combination of the longue durée of uneven capitalist development and financialisation's end, not the reason for the current turmoil in the EU. These developments are, first, deeper transformations of capitalism towards financialisation and rent-seeking activities; second, the real-estate boom-bust, at different degrees in the four countries, which acted as a bridgehead for the crisis; third, the omission of certain spatial, economic and political preconditions for an optimal currency union; fourth, the uneven intra-European terms of trade flows; and fifth, the undemocratic and authoritarian multi-scalar governance of the EU and the Eurozone, which is an elitist top-down project with a lack of mass popular support.

These deeper and highly interlinked structural, political and geographical developments did not attract attention and later proved fatal to the project of monetary union itself.

Transforming capitalism and bringing rent back in

The Maastricht Treaty and the introduction of the euro coincided with major capitalist transformations in the global North. From the late 1980s onwards, the productive sector in countries and regions exhibited slow growth – but variably between regions/countries in the EU. Profit rates remained below those of the 1960s and 1970s, unemployment became persistent and wages remained stable, and this was while productivity was rising. Geographically, these changes mark, as Dicken (2015) wrote, a "global shift" of productive activities towards the East, primarily to China and Southeast Asia, but also towards Latin America. In Europe, the shift of production towards former socialist countries, as we saw in the previous chapter, intensified after 1989 and the fall of Berlin Wall, with massive re-location of productive units mainly from developed northern-central to eastern regions but selectively from southern regions as well.

It is the period during which financialisation took the lead with the help of active intervention by the state. Deregulation of old welfare apparatuses and re-regulation for the benefit of capital (e.g. prices, quantities, provision of liquidity, cross-border capital flows, cutting welfare payments, privatisation of public utilities) plus technological innovation helped the financial sector to create new "products" and to attract the majority of surplus capital. As many authors observed (see, among many, Arrighi, 1994; Foster 2007; Harvey, 2010; Lapavitsas, 2013), the rise of finance was the result of deeper changes within capitalism, among which four seem to be more widespread. Firstly was the involvement of non-financial firms and entities in financial activities. Industrial and commercial firms, but also all kinds of institutional investors (including municipalities, universities, pension funds and hedge funds) increased their assets via financial speculations. Moreover, welfare cuts, lower wages and credit card expansion made middle- and working-class households become increasingly dependent on formal finance, and they incurred debt in order to secure access to vital goods (housing, education, health) previously provided by welfare state institutions, but also because in many cases pay stagnated and housing costs rose. Furthermore, banks and finance firms focused on transactions in open markets, making massive profits from trading future earnings by creating "securitisations", that is bundling up loans, including non-performing ones, and selling them on as assets. Finally, states, regions and municipalities, due to diminishing public revenues, began to depend more and more on bank loans or investments in securities to finance their daily operations and to build speculative mega projects. Their debt accumulated as financialisation proceeded, and in this way, private, public and municipal debt became a mechanism for capturing social wealth and political control (Lazzarato, 2012). As the French Regulation School and Lazzarato argued, it was a major shift in the "regime of accumulation", away from the Fordist productive regime towards the "regime of accumulation with financial and debt dominance" (see Aglietta, 2000; Boyer, 2000; Lazzarato, 2012).

The new regime of accumulation can also be described as a switch from the primacy of surplus value production in goods and services to surplus value appropriation via all kinds of rents (M. Hudson, 2010; Harvey, 2014; Sayer, 2015; Standing, 2016). Surplus capital was switched as fictitious capital from the "primary circuit" of investments for goods and services production to the "secondary circuit" of investments in speculative assets providing rents, including land (Harvey, 1982: 264–266; R. Hudson, 2005: 28–29). Since the late 1980s, and particularly from the 1990s onwards, the switch from productive activities to assets providing rents correlates with the dominance of neoliberalism and finance capital and in Europe with the Maastricht Treaty. Suffice to say, however, that in advanced capitalism, it is difficult to separate the financial from the productive sector, since they constitute a unity for any economic activity. For example, real estate developers cannot sell houses without finance and consumer credit. And finance is no less real than the "real economy", a commonly used distinction in the media. As Lapavitsas (2013: 798) said, the current financial transformation of capitalism has its roots "...within the fundamental relations of non-financial enterprises".

Box 3.1 Profits from rents

We were used to thinking of rent as something natural to pay for using land or a house. Classical political economists from Adam Smith and Karl Marx to contemporary radical thinkers such as Michael Hudson, David Harvey, Guy Standing and Andrew Sayer are highly critical of rent, because land exists in nature; it is not a produced commodity and therefore has no price in capitalism. What we buy and sell as land is a "fictitious commodity", as Marx and Polanyi said. So why do we pay for it? Because rent is a *distributional social relation*, not a productive one. It derives from the power of the landlord's possession, which in the first place was an act of force and grabbing. Rent is a drain on surplus created by others and therefore is *unearned income*. Radical social scientists and ecologists use the term rent to cover multiple sources of unearned income such as technological tariffs, cultural and biological patents and privatised public utility companies among others. The neoliberal and financial transformation of capitalism in particular regions of the global North prioritise "rent-seeking" activities, that is, seeking control of assets and resources that can be used to extract rent. (see more in Harvey, 1982, 2014; M. Hudson, 2010; Sayer, 2015).

In this way, as powerfully argued by Standing (2016), a new rentier economy and a new "corrupted" class of rentiers emerged in the global North, increasingly "profiting without producing" (Lapavitsas, 2013:793). This new class included not only very rich individuals, but also large institutional investors such as industrial companies (e.g. Ford, General Motors, Google, Apple); banks, finance and insurance companies; hedge and pension funds; private equity; the church and other religious institutions; universities, charitable institutions and municipalities; and individual states. It is impressive how neoliberalism changed the character and function of these institutions to make them speculative rentiers.

The main beneficiaries of speculative investments in assets are three interrelated sectors/categories. First is the finance and insurance sector, which provides intangible and innovative products. The dogmatic belief that these products have low or zero risk, as Varoufakis (2011) explained, was based on complicated mathematical models quantifying human behaviour. In turn, profit risk estimates were rated positively by international rating companies, and the whole package appeared seductive to millions of investors who trusted these products and invested their savings. The unfortunate sequel became well known, especially after the collapse of Lehman Brothers and the crisis in SE. The second sector in asset investments derives from the neoliberal dogma "less state more market", which opened the door for massive privatisation of public companies and assets. Banks provided easy loans to speculators who applied, what Harvey (2003) calls, "accumulation by dispossession". Many of those former public utility companies are natural monopolies, so in a short time their new owners became billionaires from rental income. The third sector is land, which exists also in the previous two sectors as many financial and privatised public companies include land and other buildings in their portfolio. Investments in land rents include real estate everywhere, integrated tourist development, extractivism, land grabbing (particularly in the global South and in former "socialist" countries), "mega project" construction, conventional energy projects and also renewable energy and development corridors. Finally, other sectors include conspicuous consumption, the art market, gold, diamonds and all kinds of sponsorship. Neoliberalism and financialisation are responsible for two major class shifts. The first is inter-class and concerns a shift from labour to capital, as is evident in the distribution of earnings as part of GDP. The second shift is intra-class, and it is pictured in the shift of wealth within the capitalist class, to the infamous 1 per cent. As Andrew Sayer (2015: 18) describes: "...from those whose wealth comes primarily from the production of goods and services to those who get most of their income from control of existing assets that yield rent". Much of the income of the second group is unearned, i.e. they have not worked for it.

However, the argument for the shift from production to rent extraction activities needs to be socio-spatially specific. It cannot be generalised without reference to the global shift of production towards the global South and Asia, particularly to China, where productive activities still dominate – although this is changing rapidly after the massive, large-scale and fast urbanisation there. In Europe, regional differences in productive structures should be taken into account as well as the de-localisation of industries towards former "socialist" countries in the East. Production for industrial exports is still dynamic in Germany, and this is one reason, among others, for the crisis and debt creation in the Eurozone.

A good proxy for the capitalist switch from productive to rent-seeking activities is the changes in the proportional share of (GVA) Gross Value Added of different sectors in the national GDP of each country. In the USA, the largest capitalist economy until the new millennium, the shift of investments from production of goods and services to tangible and intangible assets providing rents began in the early 1990s. As Figure 3.1 shows, this period was a turning point after which the proportionate share of the so-called "FIRE" economy (Finance, Insurance, Real Estate) surpassed that of manufacturing. By 2009, the second year of financial crisis and the real estate bubble in the USA, the FIRE economy's share of GDP was 20.5 per cent while that of manufacturing and agriculture was 13.2 per cent.

In Europe, comparative data for the ratio of GVA to GDP exist from 1995 onwards (Eurostat, 2017). In Figures 3.2–3.6, activities providing rents (A) include finance, banking, insurance, real estate and services to the above. Production (B) includes agriculture, forestry, fishing, manufacturing and producer services. Construction (C) is shown separately because of its importance in SE societies, and because the data made no distinction between construction related to real estate and for public use. The case of Germany (see Figure 3.2), the strongest EU economy, is quite different from that of the USA. Although A is constantly higher than B, both evolve in parallel and with minor differences, an indication of the strong

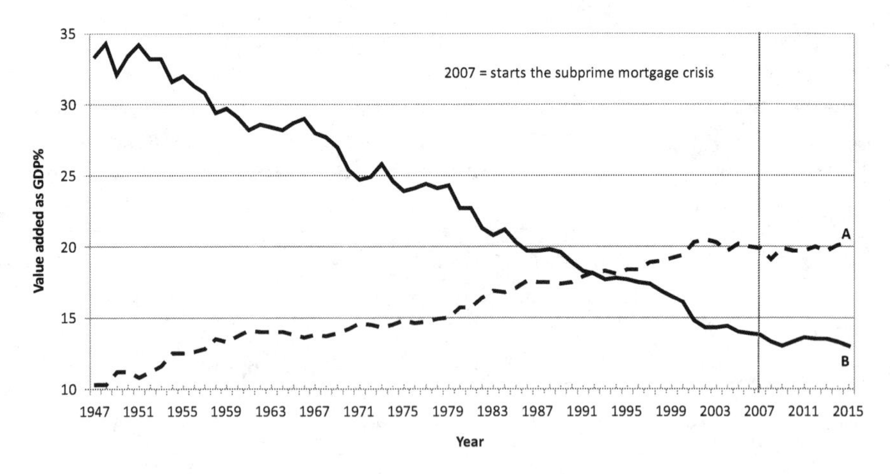

Figure 3.1 USA: Value added as percentage of GDP for selected sectors, 1945–2015.
Source: Bureau of Economic Analysis (2016).

performance of the manufacturing sector for exports and its complementary relation to finance. Of interest is also the decline of manufacturing at the beginning of the crisis in SE, i.e. fewer exports to the South, and its quick recovery after 2011.

The picture changes for the countries of SE (see Figures 3.3–3.6). All four show similarities with the US pattern, although the shift towards rent-seeking activities occurred later. In Greece, the shift happened in 1995, where both A and B had reached 18.4 per cent of GDP, while by 2015 A was 24.5 per cent and B 12 per cent of GDP. In Italy, the shift happened around the same period, in 1998, and by 2015, it had the highest ratio of FIRE activities among the SE countries, i.e. 26.1 per cent of GDP, higher than the USA. Portugal and Spain had similar performance in terms of FIRE/GDP ratio in 2015, 21.5 per cent and 21.4 per cent respectively, with Spain having a stronger construction and manufacturing sector. The shift in the Iberian countries occurred even later, i.e. in the second half of 2001 in Portugal and by the second half of 2004 in Spain. These data illustrate the discussion in the previous chapter in terms of neoliberal policies, financialisation, intensifying privatisations and the slowdown of manufacturing activities in the four countries from the early 1990s. They also highlight the gradual transformation of Southern economies before the crisis towards rent-extracting activities, particularly in real estate and finance. The process was also fuelled by the flow of surplus capital from Central-Northern economies to SE. Due to the monetary policy of the ECB, southern states were unable to dampen the speculative effects of cheap money (Clark, 2014), while wage and price inflation became permanent in the South. After the introduction of the euro, middle-income and poor households in SE felt as if they were part of the "first-class club" and sustained high levels of consumption, while their countries' productive basis was

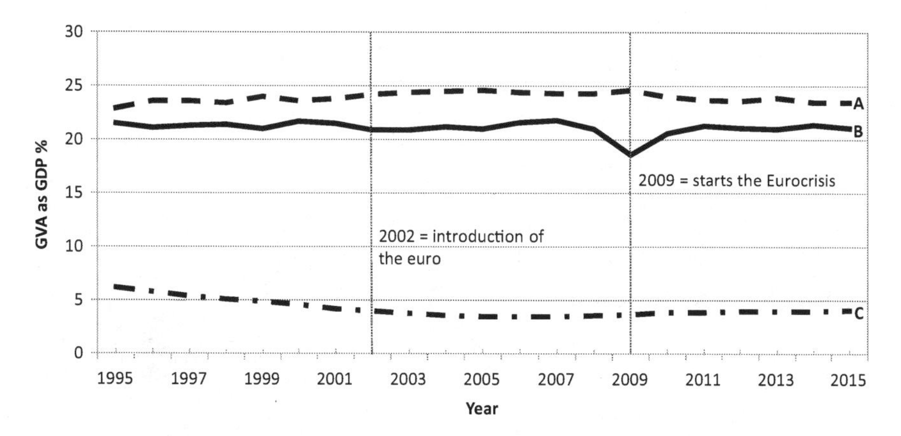

Figure 3.2 GERMANY: Gross value added as percentage to GDP for selected sectors, 1995–2015.

Source: Eurostat (2017), "Gross value added and income by A*10 industry breakdowns" in *Eurostat Database.*

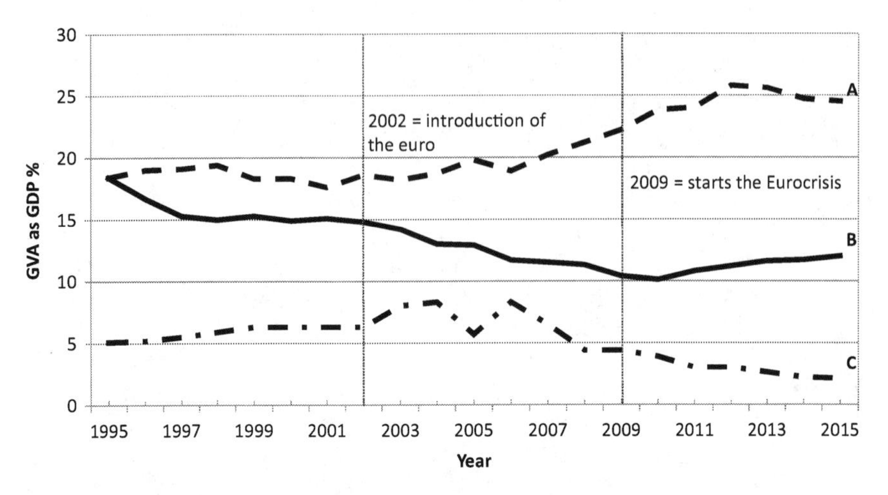

Figure 3.3 GREECE: Gross value added as percentage of GDP for selected sectors, 1995–2015.

Source: Eurostat (2017), "Gross value added and income by A*10 industry breakdowns" in *Eurostat Database*.

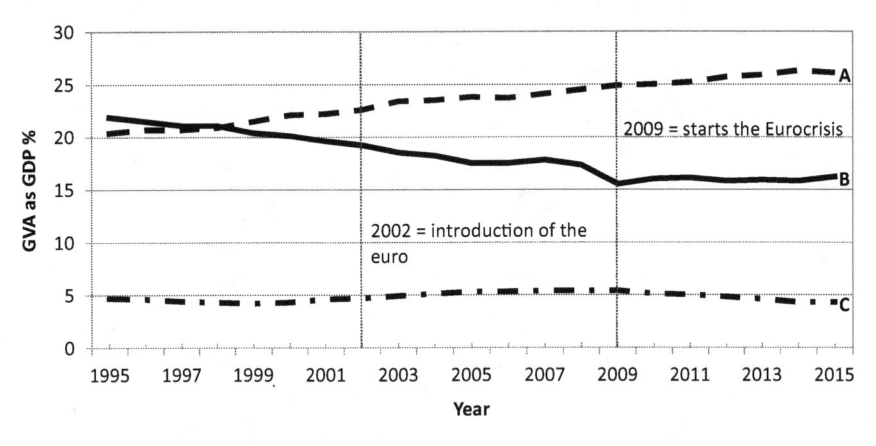

Figure 3.4 ITALY: Gross value added as percentage of GDP for selected sectors, 1995–2015.

Source: Eurostat (2017), "Gross value added and income by A*10 industry breakdowns" in *Eurostat Database*.

weakening, as shown in the four diagrams. On the one hand, "bubbles" of all kinds occurred in real estate, banking and credit-fuelled consumption. These steadily increased the fiscal problems of states because taxes on capital gains were lowered during bubble periods. On the other hand, the gradual transformation of the economy in combination with the recession after 2006–2007, generated current account deficits as a percentage of GDP. By 2008, it was −15 per cent for Greece, −12 per cent for Portugal, −9 per cent for Spain and

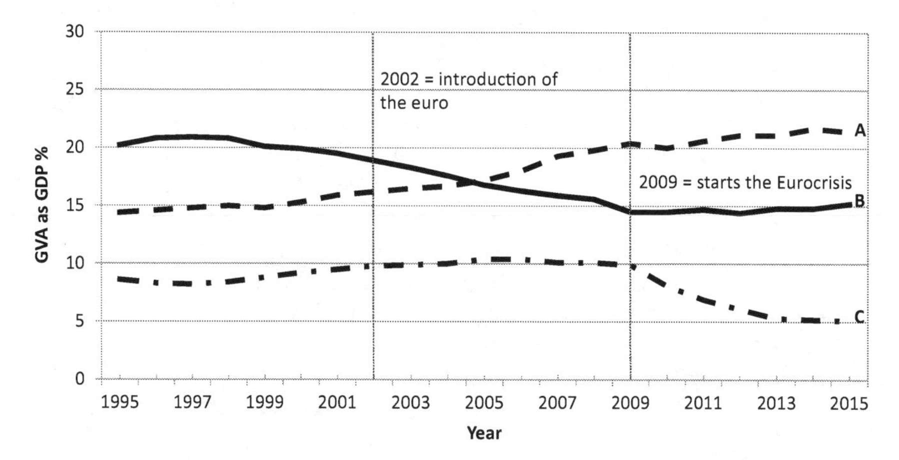

Figure 3.5 SPAIN: Gross value added as percentage of GDP for selected sectors, 1995–2015.

Source: Eurostat (2017), "Gross value added and income by A*10 industry breakdowns" in *Eurostat Database*.

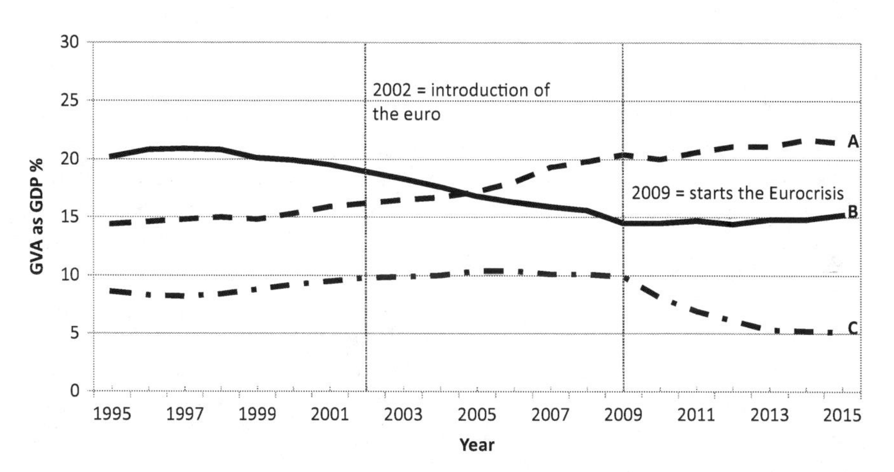

Figure 3.6 PORTUGAL: Gross value added as percentage of GDP for selected sectors, 1995–2015.

Source: Eurostat (2017), "Gross value added and income by A*10 industry breakdowns" in *Eurostat Database*.

−8 per cent for Italy, while Germany enjoyed a healthy +8.5 per cent (IMF, 2010). Deficits remained hidden during economic expansion and surfaced during recession. The immediate effect was rising state borrowing due to the decline of public revenue caused by rent-seeking activities and recession lowering the tax base, thus accumulating high private and public debt ratios to GDP after 2008 to 2009.

Financialisation, rent and real estate

Land and real estate speculation in SE is a good illustration of the capital switch to the secondary circuit of investment and rent appropriation. Furthermore, discussion of real estate is useful because economic history suggests that boom-bust cycles in real estate preceded many devastating capitalist crises, and SE is no exception (see Harvey, 2012; ECB, 2015). There is by now extensive literature on real estate bubbles and the housing crisis in SE.[1] In this discussion, however, with a few exceptions (see among them Coq-Huelva, 2013; Charnock et al., 2014) there is no reference to the role of ground rent and the changing character of Southern European capitalism (Hadjimichalis, 2014).

During the previous regimes of capital accumulation, the whole process of surplus value appropriation from land, and the corresponding urban growth, was beneficial to other sectors of the economy as well. Under financialisation, however, land and real estate are treated mainly as a financial asset when titles of ownership and shares of future income, in the form of mortgage packages, can be traded internationally. This suggests that speculation around land and rent extraction could happen without important multiplier effect to other sectors and sometimes without the material production of new built environments, although the control of fixed land properties by rentiers is required. Harvey (2014: 77) describes the process:

> The rentier class rests its power on the control of fixity even as it uses the financial powers of motion to peddle its wares internationally. How this happened in housing markets in recent times is the paradigmatic case. Ownership rights to houses in Nevada were traded all over the world to unsuspecting investors who were eventually bilked of millions as Wall Street and other financial predators enjoyed their bonuses and their ill-gotten gains.

Nevertheless, fixed construction of new houses and infrastructure must be materialised somewhere, and the development of the construction sector in SE and real estate bubbles from the 1990s onwards are indicative of the whole process. The construction sector in Spain, Italy and Greece, and to a lesser degree in Portugal, traditionally played a central role in capital accumulation. Despite processes of industrialisation in the 1970s and 1980s, domestic capital enjoyed high profits in the labour-intensive construction sector, following increasing urbanisation rates due to rural–urban migration and the demand for new housing and infrastructure. From the 1990s until 2008, the direct contribution of constructions (total value), private and public, to GDP was very high in Spain and considerable in Greece, while in Italy and Portugal it was close to the European average, as shown in Table 3.1. Since the crisis in 2013, figures declined below the EU average with the notable exception of Spain, while in Greece there was a sharp drop.

Table 3.1 The ratio of constructions to GDP, 1995–2013 (in %)

	1995	*2008*	*2013*
Spain	18.5	23.1	7.2
EU-28	7.2	6.1	5.1
Italy	7.5	5.7	5.0
Portugal	6.7	6.3	3.8
Greece	8.8	13.3	1.6

Source: Eurostat, Structural Business Statistics; IOBE (2015).

Until the 1990s, the national/regional value chain of the sector, apart from many small-medium construction firms, included several important regional and national industries in mining, cement and steel, building materials, chemicals, special transport etc., with national/regional multiplier effects providing thousands of jobs. However, after Maastricht, and particularly after the introduction of the euro, the sector faced a massive restructuring, with mergers and acquisitions, the opening of national/regional markets to large non-Southern construction companies and increasing imports of building materials from European markets, which became cheaper after the advent of the euro. Thus, the Europeanisation of this sector's value chain, despite the voluminous increase in the building stock, had minor positive local effects in the GDP of SE countries, with the exception of Spain. An indication is the reduction in the number of construction firms from 1995 to 2012: in Greece, −23.1 per cent, in Portugal, −16.8 per cent, in Spain, −14.9 per cent and in Italy, −8.2 per cent. During the same period, construction firms in Germany enjoyed an increase of 13.8 per cent and in France 27.7 per cent (Eurostat, Structural Business Statistics).

Construction of new private houses over the years resulted in high rates of home ownership. This constituted a particular SE pattern, which by 2011 was 83.6 per cent in Spain, 76 per cent in Portugal, 75.6 per cent in Greece and 72.9 per cent in Italy, in contrast with 43 per cent in Germany. The explanation of these high figures has two aspects. From an economic point of view, since the 1970s all SE countries had high inflation and fixed mortgage interest rates, so families directed savings to land properties. In addition, non-declared earnings from the informal economy had a high propensity of being invested in housing. From a social point of view, all southern countries had weak welfare systems with very limited public housing provisions with the exception of Italy. Thus, families and extended families became responsible for welfare provisions, including housing. In addition, owing to their recent peasant history, many families in SE continued to have a second property in rural or island places of origin, mainly inherited from parents or other relatives. In 2011, 23.2 per cent of all households in Greece had a second home and 14.6 per cent in Spain.[2] Finally, from a cultural point of view, across SE, owning a house remains an important status symbol plus a social/financial security.

The above pattern of home ownership is characteristic of small housing properties, usually financed by family savings and primarily exhibiting the features of use value as opposed to rental/investment value, i.e. exchange value. The 1990s, however, was a major turning point in terms of the connection of the whole process with neoliberal financialisation. The existence of surplus capital in Central-Northern European economies made it possible for Southern banks to obtain cheap money from the ECB and other European banks, predominantly from French and German banks. Regional savings banks concentrated the greater part of their investments in the real estate sector, not only lending to developers, builders and buyers, but also even becoming developers themselves (Sevilla-Buitrago, 2015b). In the four SE countries, investments in transport infrastructures, co-financed by EU Structural Funds, helped the real estate market to take off by providing access to empty lands, particularly along coastal areas, for tourist/second homes real estate and in the periphery of large cities. The whole process fuelled the switch from primary to secondary circuits of capital in SE. Marisol García (2010: 967) argues that the growth and crisis of the Spanish urban growth model, based on speculative real estate, reveals "...a particular interaction of globalizing forces with national and local processes, characterized both by specific structures of economic incentives and path-dependent cultural traits". Her statement applies to the entire SE, with differences depending on the levels of financialisation.

In Portugal and Greece, the Central Bank's regulations prevented the banking sector from developing subprime mortgages, despite several neoliberal modernisations after the 1990s, although private banks were engaged in competition to provide credit to both developers and consumers. In Spain, however, from the 2000s, mortgage-covered bonds were issued by many financial institutions, the famous "cédulas hipotecarias", which became the largest traded asset in Spain and one of the largest in the European bond market (Mayayo, 2007). In Italy also, financialisation went deeper with the Law 431 of December 1998. Financial tools, such as securitisation of loans were introduced and were purchased by pension funds, hedge funds, insurance companies, large corporations, local governments and foreign states. Consequently, in the last two decades rents in Italy have increased 130 per cent for renewed contracts and 150 per cent for new contracts. The situation is dramatic when looking at mortgages. A wave of foreclosures impacting 46,000 families took place in 2012 (Dinoto, 2013; Gibelli, 2015).

After the advent of the euro, low interest rates and the inflow of Northern European surplus capital to Southern banks in the form of loans made access to housing loans very attractive to both developers and consumers.

The average annual increase of credit to the housing buyer for the period from 1997 to –2008 was 20 per cent in Spain, 12 per cent in Portugal and 8 per cent in Greece, which together with high unemployment rates explains the far-reaching household debt and the increase in non-performing loans, the so-called "red loans", in bank portfolios. The credit access for large population segments, including the working class and immigrants, fuelled the expansion of real estate dynamics but resulted in a sharp increase in household debt. By 2012, the rate of households defaulting on repayment of housing loans to total households in Spain was 64.2 per cent, in Italy 56.1 per cent, in Greece 41.6 per cent and in Portugal 39.3 per cent, compared to Germany with only 20.3 per cent. Dominant explanations by mainstream media accused "irresponsible" citizens for borrowing too much money from banks. The reality, however, has been that banks in SE, particularly in Spain, did not take any guarantee after the inflow of cheap money from the North and offered loans even knowing that some people could not repay them. Therefore, as Coq-Huelva (2013) argues, the real estate expansion is responsible for an aggressive income redistribution from social classes relying on their hard everyday labour, to the rentier social class who enjoys unearned incomes managing the real estate business. In the entire SE, as Dinoto (2013: 43) states, "what has happened is a real paradigm shift in the model of wealth production". Active state intervention through planning laws is a prerequisite for what has been described so far (see Box 3.2).

Box 3.2 Planning at the service of real estate

At the regional and urban scales, neoliberalism promotes regional/urban entrepreneurialism and competition among cities and regions. Autonomous regions in Spain had the legal power to implement planning and real estate legislation, and, after 1986, these activities intensified across Spain with the exception of Andalusia region, which left the housing and real estate market unregulated. In the 1990s, the central Spanish state tried to reform land regulation and unplanned urbanisation by introducing two opposing legal frameworks: the Urbanism and Land Act (1990) and the Land Regime and Rating Act (1998). According to Coq-Huelva, (2013) with the later act "...all lands were considered urban (if the infrastructure was already built) or potentially urban (if the infrastructure was not built or only partially built)". The latter fuelled speculative tourism real estate.

This legislation came closer to the Greek pre-existing legal framework that allowed construction everywhere (apart from land that is regulated or protected otherwise), if the owner had 0.4 ha. In the 1990s and 2000s in Greece, several new planning laws were issued, particularly for tourist real estate and luxury urban expansions and mega projects, influenced by the

(*Continued*)

2004 Olympics. New laws introduced "several exceptions" to existing regulations, increasing the plot ratio coefficient, i.e. how much you can build on a particular plot. During the same period, large Church and monastery properties entered the market as preferential targets for foreign tourist investments (Hadjimichalis, 2014).

Portugal mega events, such as the World Expo in Lisbon and the EURO football championship in 2004, substantially increased real estate pressures and required new legislation in 1998 supporting Local Spatial Plans and New Urban Policy. Portugal was one of the first countries to introduce the Golden Visa for non-EU citizens when the purchased property was above 500,000 euros in value. Several thousand visas were given, not always with "clear" procedures. In Italy, planning for real estate takes place at the level of the 8,000 communes, which were preoccupied after the 2000s with institutional investors who looked for historical buildings and luxury properties. In parallel, real estate scandals involving illegal construction continued in Sicily, Campania and Basilicata.

In Portugal, after the 1990s, housing costs and household indebtedness increased parallel to the increase of the credit available. These contributed to an excessive increase in construction activity, particularly in Lisbon and Oporto, and the progressive abandonment and degradation of city centres. This promoted the loss of the vitality of these cities and the rise in real estate activity (Silva, 2013). The real estate champion, however, is Spain. Between 1998 and 2008 more than 6.5 million new houses and apartments were built in Spain, more than those built in Germany, the UK, Belgium and France put together in the same period. This extraordinary urbanisation is explained by high levels of profitability in real estate, between 30 per cent–46 per cent, and was the major reason for the capital shift towards the secondary circuit.

Several social actors including the banks, insurance companies, real estate developers, the state and regional governments and local politicians in municipal councils regulated the finance-real estate-political regime. Municipalities found in city sprawl and tourist real estate an opportunity to increase their tax revenues. This regime remained highly unstable due to continuous corruption and scandals at various governance levels, with the participation of multiple social actors. Box 3.3 spotlights a few examples.

In Greece, the catastrophic decision to host the 2004 Olympic Games added €20 billion to the public debt and introduced pressures for a real estate bubble. In Spain, the bubble in its real estate sector, with thousands of unsold units, generated immense problems for local banks and for other entities in Germany, France and Italy. These debt-fuelled investments were typical of the switch to the secondary circuit and have absorbed billions and billions of euros during the past two decades – far more than industry – thus becoming a prime driver of the debt crisis. A working paper by ECB (2015)

Box 3.3 Real estate scandals

Corruption scandals involving banks, developers, landowners and local and national politicians are endemic in SE real estate. The Marbella scandal on Spain's southern coast is perhaps indicative of the tourist real estate fever and concerns the illegal construction of tens of thousands of homes in the 1980s and 1990s on rural land, illegally disposed by developers after being declared by the local municipality as "land for urban development". After the scandal in 2005, thousands of people lost their holiday homes, including working class expatriates from the UK, Ireland and Germany, while dozens of socialist local politicians and developers are now in prison. In the 1990s in Portugal's Algarve coast, thousands of hectares of environmentally protected land became tourist development zones for luxury golf estates after changes in regional planning and land use legislation by the then socialist government. "Big names" like former Prime Minister José Sócrates and former EU Commission boss José Manuel Barroso were involved in these legislative changes. Furthermore, involvement in real estate business and land devaluation in Portugal after the crisis in 2011 was also responsible for the collapse of Banco Espírito Santo, Portugal's largest bank, in 2011. In Athens, Greece, the largest mall in town was built without a legal permit, using the "exceptional" legislation for the 2004 Olympic Games. The Mall belongs to the Latsis clan, a powerful ship-owning family with strong political connections. The Orthodox Church and many monasteries, among the largest landowners in Greece, were involved in several scandals as well. During 2005–2008, a major scandal, which reached the courts, was with Vatopedi monastery on Mount Athos using fake documents to illegally exchange 3,000 ha, including a lake with urban land near Athens, thus increasing its market value by four times. Several monks, an ex-minister, lawyers and a notary public went on trial.

verifies the above comments and argues that boom-bust real estate prices in SE and other countries, such as Ireland and the USA, including US states such as California, precede major economic crises – an argument advanced also by Harvey (2010) – and should be taken as a warning for what might happen next. Figure 3.7 shows the boom-bust cycle (2002–2014) of property prices in Greece, Spain, Italy and Ireland, the slow increase in Portugal and the almost flat performance of property prices in Germany. With the exception of Portugal, all other SE countries, plus Ireland, had a debt-driven real estate boom during the period from 2005–2007, preceding their severe economic crisis in 2009.

In summary, real estate boom-bust in SE contributed severely to the public and private debt crisis, first, because many firms, households, banks and public/municipal institutions were related to and/or financially dependent

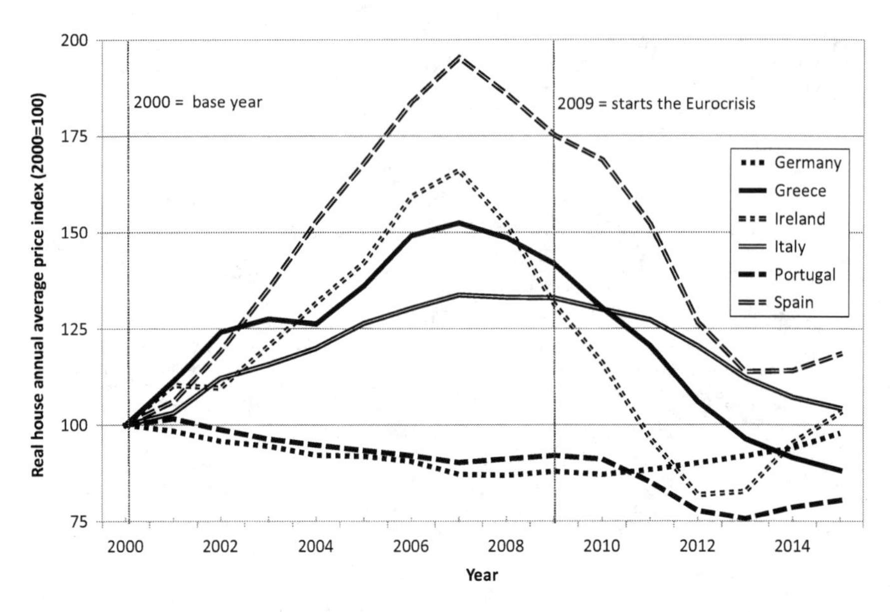

Figure 3.7 Real house price index for Germany, Greece, Ireland, Italy, Portugal and Spain, 2000–2015 (Index, 2010 = 100 rescaled to 2000 = 100)*.

Source: Eurostat (2016), "House price index, deflated-annual data" *in Eurostat Database*.

on this sector. With the widespread downturn in property prices, all were and are affected, albeit differently, with the weakest, i.e. the working class households, paying the highest price, losing both jobs and houses (for the case of Spain see, López Hernántes and Rodríguez López 2010). Moreover, real estate projects and house purchases were credit-financed with loans, often traded internationally. A default in a particular place – hence the geographical importance of the whole process – initiates a downturn domino effect impacting other sectors in other places with the well-known devastating socio-economic and political results. And finally, illusions about the continuation of existing boom trends, via the extrapolation of past real estate prices into the future, were based on the economists' belief in a strong euro and the security of the European banking system. It took only six months in 2009 for the collapse of this financial illusion.

Transformed European capitalism meets an infant currency without a unified political space

In the early 1990s, the new regulations introduced by the Stability and Growth Pact of the Maastricht Treaty provided the framework for the functioning of the Eurozone. The above framework represents a degree of loss of national sovereignty for Eurozone states, trading national for allegedly supra-national sovereignty, and the loss is more severe for southern states (Lapavitsas et al., 2012). In the 1990s, governments in SE began the implementation of austerity programmes, long before the debt crisis hit them, to reduce the borrowing costs in international markets. Austerity, as expected, "...has compressed the only element of aggregate demand, namely public spending" (Lapavitsas et al., 2012: 35) plus consumer spending.

The unification of Germany in the same period reinforced the widespread perception that German hegemony was a risk and that a solution should be found to prevent major economic and political imbalances among the original founding members of the EU. France and Italy wanted the euro for sovereignty purposes, but as many agree, Germany accepted the idea only on condition that the new monetary unit would operate in a German ordoliberal way. Working against theory and history, the EU failed to realise the efficiency gains associated with fiscal integration, something realised after the debt crisis. According to Stockhammer (2014), there are analysts who believe that implementing the monetary integration before the political integration was a concession to German elites; others argue that it was a conscious strategy by the German Bundesbank, which wrote the rules for the new currency. In both explanations, the winners are the German elites in the export sectors, mainly in Southern Germany and some regions in the North.

To prepare for the introduction of the euro, the (ERM) Exchange Rate Mechanism and the EMU were established in the 1980s. This was a German concern favoured also by its "satellites", i.e. the Netherlands, Belgium, Austria and Scandinavia, especially Finland. This caused a geographical division of Europe into four groups based on economic strength and positive trade balance (see also Burroni, 2016). In the first group, having both indicators strong, belong particular regions in Germany and the "satellites" (e.g. Baden-Württemberg, Bavaria, Lower Austria). In the second group belong regions in France that are economically strong but erratic in terms of exports. Italian regions in the centre-north and some regions in Spain (e.g. Catalonia, the Basque Country, Valencia and Madrid) belong in the third group, being less strong economically, but becoming quite positive in export performance when the lira and the peseta were devalued. The majority of regions in Spain, Greece, Southern Italy and Portugal belong to the fourth group, having both weak economic structures and weak export performance (Hadjimichalis, 2011; Bellofiore, 2013; Simonazzi et al., 2013; Marques, 2015).

The major problem in the very formation of the Eurozone is the so-called "national convergence criteria" (price stability, low interest rates, stable exchange rates and limits on the size of budget deficits and national debt) and the neglect of spatial or regional convergence. In the debate on the euro in the 1990s and early 2000s, and later on during the debt crisis, very little attention was paid to geography in relation to the four important conditions for a successful monetary union. According to Magnifico (1973), Thirwall (2000), Martin (2000) and Hadjimichalis and Hudson (2014), four necessary spatial and economic conditions for an optimal currency area have been identified:

a Regional economic and social structures should have a degree of economic similarity and relative equality in the value of export and import flows in order to avoid trade surpluses in export regions and trade deficits in importing regions. In the absence of such similarities, neoliberal restrictive monetary policies will deepen geographically uneven employment/unemployment.
b Regional economic and social structures should have high rates of geographical mobility for capital and labour. If such mobility is weak, especially in regards to labour,[3] as was and remains the case within the EU, cyclical crises may lead to persistent regional inequalities. An important parameter here is the direction of labour flow.
c Regions should have a similar propensity for inflation.
d There should be an automatic fiscal mechanism that, through a centrally organized tax and benefits system, will compensate for different national and regional shocks and growth rates. There should also be a central bank to operate as the "lender of last resort". This last condition is absolutely critical.

Crucially, none of these conditions existed at the time of the euro's introduction, nor do they today. The Eurozone does not meet the above criteria of an optimal currency area. It is an economic *and* political project without spatial considerations, an extension of the EU commitment to the single market while ignoring that markets are always geographically determined. Already in 1993 and after the crisis in 2010, Krugman (1993; 2010) illustrated some of the difficulties that the euro would face at the regional level, criticising the introduction of only national criteria to join the Eurozone. He also suggested that the geographical mobility of capital and labour needed to be consistent with regional and unified national social welfare, something in operation in the USA and missing, intentionally I could argue, in the EU. Along the same lines, Pivetti (1998) argued that the major problem with the euro's architecture is monetary integration without a joint authority over a joint budget and a joint balance of payments. Taking into account the highly uneven development among countries and regions of the union, this omission would eventually work against cohesion.

Since the formation of the Eurozone, only capital mobility has improved slightly. Labour mobility has remained low, while regional divergence in economic structures, employment/unemployment, inflation and welfare have accelerated. Thus, SE regions, and particularly those with fragile productive structures, have become the weak link in a very unstable monetary union. As Martin (2000: 9) argues: "When countries form a currency union, their regions in effect become 'twice removed' from monetary policy and control", because member states transfer such policies upwards to the monetary union. Without a centrally organised tax/benefit system, and under the false assumption that sectoral and regional imbalances would be self-corrected by markets, the Eurozone moved towards monetary integration. Very few anticipated that a further geographical concentration of economic activities in fewer regions and a deepening of unevenness between people and places could be expected, in line with the historical example of the Italian Risorgimento and the imposed monetary union at the expense of banks, industries and particularly workers in the Italian South (Del Monte and Giannola, 1978). And even fewer raised the likelihood of a capitalist crisis (what neoliberals call asymmetric shocks) unevenly affecting Eurozone regions and member states in the absence of nominal exchange rate instruments and a central redistributive mechanism (Flassbeck, 2010).

In the European Parliament, however, voices from the European Left Party and the Greens argued in 2005 (see *Avgi*, 23 May 2005 [in Greek]) that a new regulatory system is needed. For the transition towards a common currency and from tax competition to tax cooperation, a further increase in the EU budget redistribution is needed, which could promote solidarity by narrowing uneven geographical development. These voices were not heard, and a new version of the "old" division of labour between the Central-North core (mainly German regions, Austria and the Netherlands) and the Southern periphery emerged, challenging prospects of convergence. This division of labour also includes Ireland, which followed a different developmental path.[4]

Uneven development and uneven trade flows, not only debt, tell the story of the crisis

The entrance of southern regions into the multi-scalar, unaccountable and uneven Eurozone space, plus their own endogenous economic weaknesses, resulted in a progressive loss of competitiveness. Figure 3.8 depicts this loss of competitiveness at the national scale. Whereas Germany increased its competitive position after the introduction of the euro and remained steadily high after the beginning of the crisis, the competitiveness of SE countries after 2004–2005 decreased sharply, with Greece leading the tendency to fall.

Noteworthy is also the continuation of poor performance after a decade of internal devaluation at the expense of labour, with Spain showing a slight improvement. This is a clear indication of the inadequate combination of painful adjustments imposed by the Troika and the one-programme-fits-all policy made without consideration of the structural characteristics in each social formation.

The articulation between national and regional scales needs further discussion. Aggregate data, e.g. trade balance or export figures, prioritise the

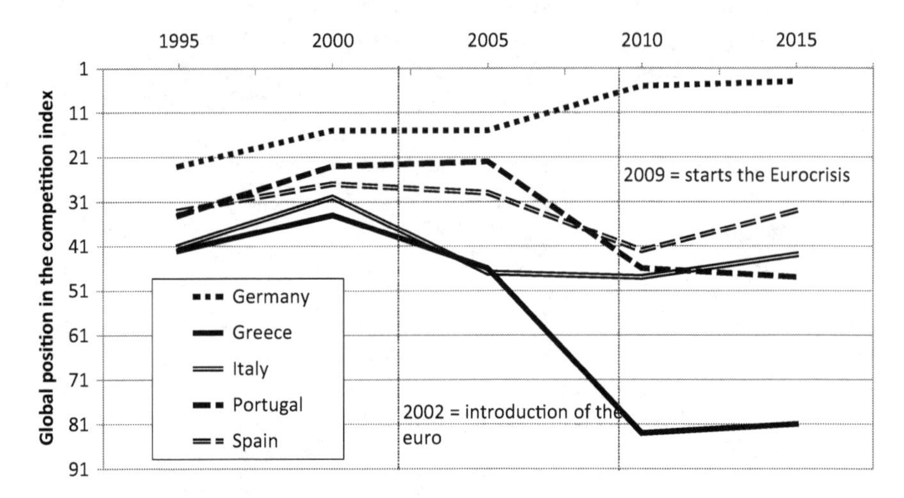

Figure 3.8 Competitiveness of Germany, Italy, Spain, Portugal and Greece, 1995–2015.
Source: World Economic Forum (2016), "The Global Competitiveness Index Historical Dataset" in *World Economic Forum-Competitiveness Rankings*.

dynamics of uneven national relations at the expense of uneven regional ones. But this is a statistical representation that masks particular inter-firm relations that operate in different spatial environments and under particular capital–labour relations, the euro being the only equal parameter. Furthermore, these aggregate data hide socio-spatial divisions of labour and unequal class relations within firms and regions. Although it is difficult to highlight the above at any analytical step, it is crucial to keep it in mind in order to avoid reducing socially and spatially embedded actors to notions of "Germany", "Greece" and "Ireland", or to "core" versus "periphery".

In their otherwise brilliant study, Lapavitsas et al. (2012) adopt an old-fashioned core-periphery model in which the core dictates to the periphery its role and function, and the periphery remains dependent. For Lapavitsas et al., this is a sound explanation for deficits in the periphery and surpluses in the core. This analysis takes for granted the passivity of social actors in the South: politicians, elites, firms and citizens alike. The macro-economic framework used by Lapavitsas et al. at the national scale fails to recognise events and actions at other scales, including the strategic decisions by firms (e.g. de-localisation), the role of the family and SMEs, as analysed in previous chapters. In other words, they fail to recognise the importance of the local and the dialectical, as well as the contradictory interplay of endogenous/exogenous factors and the decisions by local social actors, who acted, of course, under conditions not of their choice.

Although the monetary union increased the intra-euro area trade by between 12 per cent and 30 per cent over a five-year period, southern regions benefited much less. This was because the market access improvements benefited firms in some North-Central European regions rather more than those in SE regions, as a result of five main factors. First, the absolute reduction in trade cost for the entire Eurozone increased the relative disadvantage of those regions that were behind at the time of monetary unification, particularly SE regions that entered the union with higher nominal exchange rates and were located far away from Europe's core. Second, the loss of the old nation-state regulatory framework, which protected Southern firms via currency devaluation, bilateral international trade agreements and provisions of investment incentives, shifted competitiveness within the Eurozone to unit labour cost. Third, the German export-led strategy resulted in an import-led outcome for southern regions that did not achieve a sufficient level of diversification and specialisation in their productive structures. Fourth, the increasing integration of central-eastern regions within the supply chain of German industries, as Simonazzi et al. (2013) argue, "...has gone hand in hand with an impoverishment of the productive matrix of those southern regions less connected with Germany" (p. 664). And fifth, the operation of market forces drew activity and channelled exchange value flows through trade, generating surpluses for northern-central regions at the expense of southern regions.

These points need further scrutiny. Although classical political economists give particular attention to trade as a main source of unevenness across sectors, firms and space, recent theoretical accounts tend to neglect it. Figure 3.9 shows the increase in value of German exports to major world areas between 1995 and 2007. In terms of both value and increase, SE sits at the top, with the bulk of products directed to the larger markets of Spain and Italy.

The diagram also shows how important SE has been for German exports since the introduction of the euro, when exports to the USA slowed.[5] Bellofiore (2013: 505) explains trade relationships between SE and Germany:

> ...Germany, like the rest of northern Europe, had a historical need to export to SE, where it realised the largest part of its profits. Thus trade deficits in France, Italy, Spain, Portugal and Greece were crucial to Germany's competitiveness. They also held down the nominal valuation of Germany' currency, the euro....This structural strength is due to Germany's specialisation in technology sectors, advanced machinery and high-quality manufacturing, and not just wage deflation.

Bellofiore's argument answers in part the question why Germany accepted into the Eurozone the problematic and weaker Southern economies. During the same period, following Figure 3.10, there was a continuous negative trade balance for the four SE countries vis-à-vis Germany, with only Italy showing a positive balance in 1995. The negative trade balance was higher for Italy and Spain due to their larger economic size, and for the four countries, it was continuous until 2015, an additional indication of the inadequacies of adjustment policies. The policy of internal devaluation to increase

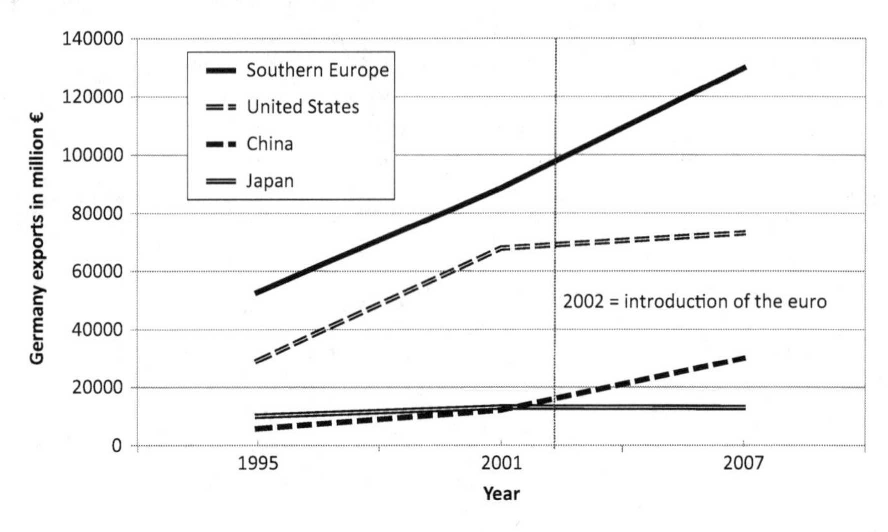

Figure 3.9 Germany's value of exports to Southern Europe, the USA, China and Japan, 1995–2007 (Southern Europe refers to Greece, Italy, Portugal and Spain).
Source: Eurostat (2011), *External and Intra-European Union Trade-Statistical Yearbook*.

competitiveness and exports ignores the impossible mission of a universal replication of export-led growth. Germany has been able to export because other SE economies have been doing the opposite.

Unequal trade is a basic condition for uneven capitalist geographical development that frames – and is framed by – the production of commodities, the geographical circulation of surplus value embodied in those commodities and the spatially differentiated realisation of surplus value (Hadjimichalis, 1987; R. Hudson, 2001). Southern countries, with the exception of Italy and some Spanish sectors, are characterised by low product specialisation and low capital and technological intensity. Thus, the realisation of surplus value by SE commodities is less profitable, compared to the same process for commodities from Germany having higher technological intensity (higher capital composition), fuelling further the interrelated surplus-deficit polarisation in the EU.

The low performance of exports from SE countries mirrors their narrow productive base, particularly in Greece and Portugal, and their inability to compete with cheaper imports (Weeks, 2014). This again underlines the importance of labour productivity and nominal labour cost as major components in competitiveness, given the very low geographical mobility of labour between EU regions. Between 1995 and 2007, Greece's net exports (65 per cent of which were to other EU countries) were sluggish, but domestic demand rose at a healthy 2.3 per cent, as imports from EU countries became "cheap", owing to the euro (Eurostat, 2006, 2009). Real compensation to labour increased at 1.9 per cent per employee annually, a little less than productivity per employee, which increased 2.1 per cent annually (INE/GSEE, 2010). Nominal

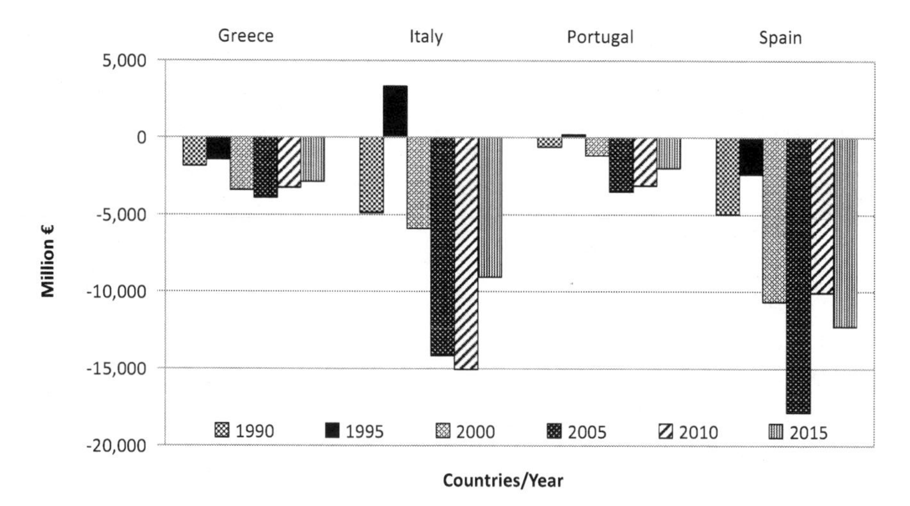

Figure 3.10 Trade balance with Germany: Greece, Italy, Portugal and Spain, 1995–2015.

Source: Eurostat, *External and Intra-European Union Trade-Statistical Yearbook,* 2016.

unit labour costs (shown in Figure 3.11), an important measure of competitiveness between members of a currency union, advanced at a rate of 2.8 per cent per year and in 2009 reached a level of 145 for Greece, 133 for Italy, 135 for Spain and 125 for Portugal, from a base of 100 in 2000.

Labour productivity for SE (with the exception of Spain) increased as well, with Greece reaching 140 in 2009 (where 1995 = 100), Portugal 127, Germany 120 and Spain 109 in 2009 (OECD, 2010). During the same period, German firms and the German state accumulated a huge current account surplus, culminating at 8.0 per cent in 2007. Germany's net exports to SE exploded between 2000 and 2009, but domestic demand stagnated at a 0.2 per cent annual increase.

German elites took advantage of the unification with the former Eastern Germany in 1989 by keeping increases in compensation per employee below those in real GDP per person employed. Further labour market reform in Germany during the Schröder administration, known as the Hartz reform, around 2003, reduced workers' pay and introduced high flexibilisation. Another geographical strategy of German capital to reduce labour cost, beginning in the late 1980s, was the de-localisation of simple manufacturing tasks to Eastern Europe, mainly to regions with an industrial tradition and/or in former "Kombinaten" areas.[6] This helped to sustain high productivity in manufacturing for export while contributing to a sharp fall in Germany's unit labour cost. Together with the locking in of other EU countries to fixed exchange rates via the EMU, and later with the euro, this strategy of socially and spatially differentiated "triple" labour markets (former Western and Eastern Germany plus

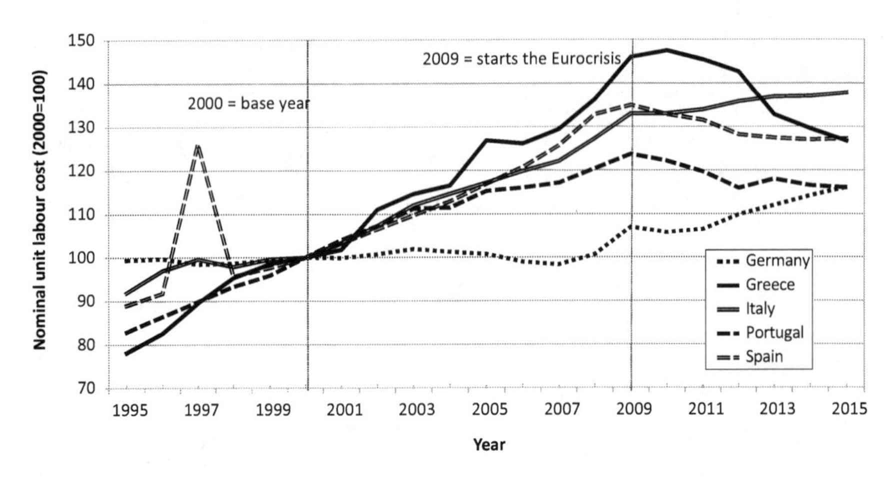

Figure 3.11 Nominal unit labour costs in Germany, Greece, Italy, Spain and Portugal, 1995–2015*.

Source: Eurostat "Labour productivity and unit labour costs" in *Eurostat database,* 2016.

*Nominal unit labour costs based on persons (Index, 2010 = 100 rescaled to 2000 = 100).

neighbouring countries) contributed decisively to the export success of Germany after 2000. The result, however, apart from shrinking internal demand and increasing exports, was an impoverishment of the German working class; in 2010 more than 15 per cent of the German population was affected by poverty.

Thus, as shown in Figure 3.11, nominal unit labour costs rose only marginally in Germany in the decade from 2000–2010 (at the expense of the working class, particularly in eastern regions), reaching a level of 108 in 2009 from a base of 100 in 2000. This simply means that the production of a comparable good or service that was produced at the same cost in 2000 in all member states of the Eurozone, and could be sold at the same price, then cost 36 per cent more if it came from Greece than if it were produced in Germany (Papic et al., 2010). The difference for Italy was 27 per cent, for Spain 25 per cent and for Portugal 16 per cent. In this respect, the euro provides considerable economic and political benefits to German capital (and to a lesser degree to French, Benelux, Nordic and Austrian capital). Germany's neighbours were unable to undercut German exports with currency depreciation, and in turn German exports gained in terms of overall Eurozone exports to both global and Eurozone markets. Thus, the SE's external deficit reflected, among other things, the strategy of German and other European elites, which is aimed at regaining market shares and political hegemony in the Eurozone by squeezing production costs through the freezing of compensation to workers, while keeping high the intensity in technology and innovation (Medelfart et al., 2003; Jabko, 2010). The combined effect was an acceleration of the destruction of productive structures in SE regions, including many prosperous intermediate regions – among them some of Third Italy's famous industrial districts. This process had started earlier in the 1990s, as discussed in the previous chapter (see also Hadjimichalis, 2006). The widening trade gaps went from problem to disaster when the euro was introduced, and this became visible in former competitive export sectors and regions. Box 3.4 describes the case of the footwear industry.

After the euro, financialisation in SE was driven by financial flows from Central-Northern banks. In the uneven trade relations, Germany benefited from weak productive structures in SE, but also from the euro's adoption by Southern economies, because the D-mark would be stronger and Southern currencies weaker, thus making German exports more expensive.

Box 3.4 Crisis in the Southern European footwear industry

Footwear is of little importance in EU industry as a whole (0.6 per cent of total manufacture, 2003), but for several SE regions, it is a crucial part of the local economy and society. Italy has a prominent position as the greater producer, contributing 43 per cent of the total value for the EU 15, followed by Spain 17.6 per cent, Portugal 10.3 per cent, Germany 9.8 per cent, France 9.3 per cent and Greece 4.2 per cent (Eurostat, 2001). Portugal presents the highest industrial specialisation with 4 per cent share of footwear in its total manufacturing activity, Italy 2 per cent, Spain 1.3 per cent and Greece 0.5 per cent.

From the early 1990s onwards, the footwear sector faced a considerable crisis due to the opening of other, more competitive production locations. Extreme competition from Asian, North African and Eastern European countries, rising costs in SE, the removal of various protectionist measures and the introduction of the euro, all made SE footwear very expensive. From 1995–2005, European footwear production declined from 1.1 billion pairs to 700 million, exports declined by 36.3 per cent, while imports (mainly from China, India, Thailand and Eastern Europe) increased from 300 million to 1.25 billion pairs (EU Confederation of the Footwear Industry EU-CFI, 2008). A special feature of the entire footwear *filière* is its spatial organisation. With the exception of Greece, in all other southern countries, industrial production continues today to take place in historical industrial districts, where externalities and agglomeration economies continue to provide some advantages to small and medium enterprises. Historical traditions, local labour market characteristics and local knowledge remain important location factors. For Italy, production is concentrated in industrial districts in the Marche, Emilia-Romagna, Tuscany, Veneto and Lombardy, in Spain in Valencia and La Rioja, in Portugal in the North, mainly in the hinterland of Oporto and in Greece mainly in Western Attica: a "diffused industrialisation" pattern.

(*Continued*)

From the four southern countries, only Italy has a substantial number of large companies (more than 250 employees/firm and more than 70,000 euros sales value/firm per annum), while Portugal has three to four very large companies with more than 1,200 employees. The absolute majority of Southern firms in the footwear industry are small or very small companies with 75 per cent of all firms having less than 10 workers, as an average over the four countries. These mini-companies are functional and essential to the whole filière and work primarily as subcontractors to national and other European large firms (from France, Germany, England, Austria, Denmark and Finland). For them, remaining competitive means dependence on highly skilled and low-wage labour. The Marche in Italy, Valencia in Spain, Norte Portugal in Portugal and Western Attica in Greece account for the majority of footwear firms and jobs in their respective national economies, as table 3.2 shows, with Norte Portugal having the highest figures. From the table, one can also see important negative developments from 1991 to 2005 in terms of the reduction in firms and employment, with Western Attica showing the highest negative figures, followed by the Marche.

The four regions have other important differences. First, they are highly unequal in terms of actual production, product quality and export orientation; second, in terms of reputation, brand name and export performance; third they differ in de-localisation patterns. The Marche leads the group, followed by Valencia, Norte Portugal and Western Attica (Blim, 1989; ARMAL, 2003; Ybarra et al., 2004; Hadjimichalis, 2006; Belso-Martínez, 2010).

Table 3.2 Regional clustering of footwear firms and employment and crisis (different years)

	% of firms country = 100	*% of employment country = 100*	*% reduction of firms*	*% reduction of employment*
The Marche (2002)	51.1	60.0	−5.6% (1995–2003)	−13.0% (1995–2003)
Valencia (2003)	68.2	65.4	−9.3% (1991–2004)	−12.3% (1991–2004)
Norte Portugal (2003)	82.1	92.2	−3.6% (1991–2004)	−9.8% (1991–2004)
Western Attica (2005)	65.2	71.3	−10.1% (1991–2005)	−19.1% (1991–2005)

Source: various local footwear associations, fieldwork by the author.

In this respect, the spatial dimension of financialisation benefited sectors and regions unevenly across the Eurozone and was more damaging in SE due to the overall weakness of the productive structures.

Taking together unequal trade relations, the destruction of SME-based regional productive structures, serious problems from debt-fuelled restructuring towards financialisation, real estate and mega projects, plus the wider recession imposed by neoliberal policies during 2004–2009, the tax basis of public revenues in southern states was challenged. Along with their chronic class priorities and inefficiencies in collecting taxes, corruption, clientelism and bad planning, SE states faced a dramatic decline in tax revenues. Budget deficits increased, and SE states were forced to borrow money from the European financial market, increasing their public debt further, which then went above the 3 per cent of the GDP limit imposed by Eurozone regulation.[7] Figure 3.12 shows the increase in the fiscal balance in all SE countries (with Greece leading the race again), but also that Germany too passed this limit in 2009 when the debt crisis started. Public finance was in trouble in the entire Eurozone. Low interest rates in the Eurozone allowed SE banks to obtain loans cheaply from German, French, Belgian and Dutch banks so that their exposure by 2008–2009 had increased considerably. When the US financial crisis of 2007–2009 arrived in Europe, southern states, on top of the recession and their shattered regional productive structures, spent billions of euros to support their banks, which had been overexposed through real estate financing.

Supporting the banks reduced public revenues even further and combined with the recession caused the ratio of public debt to GDP to skyrocket. International rating agencies, the "gate keepers" of neoliberal orthodoxy, rated Greek state bonds in May 2010 to "junk" status and similar negative ratings followed later for Irish and Portuguese bonds. The need to borrow money increased at the "wrong time", when spreads[8] were at the highest level: the banks that were saved by states in 2007–2009 now bite the hand that fed them.

During the period from 1995–2007, the ratio of public debt to GDP remained stable in the four countries, as high as 100 per cent–120 per cent in Greece and Italy and lower in Portugal, around 65 per cent, while in Spain ever lower at 38 per cent in 2007 (see Figure 3.13). In Greece, the structure of debt was due to the government's borrowing from banks. In the other three countries it was mainly due to the private sector, resulting in highly indebted households and corporations (Papadopoulou and Sakelaridis, 2012). Until the introduction of the euro, debt was related to domestic financialisation, a situation that changed completely after the euro to become external debt to major European banks, mainly in Germany and France (Lapavitsas et al., 2012).

After 2007, public debt increased dramatically, in Greece reaching 172.1 per cent in 2011, in Portugal, 111.4 per cent, in Spain, 69.5 per cent and modestly in Italy remaining, however, as high as 116.5 per cent. By 2015, after 10 years of austerity and five years of the Troika's "rescue" programmes that fuelled recession and negative or very slow growth, public debt continues to increase, due to GDP reduction reaching 177.4 per cent in Greece, 132.3 per cent in Italy, 129 per cent in Portugal and 99.8 per cent in Spain. By 2015, Greece's debt amounts to 311.6 billion euros, Italy's 2,172.6, Portugal's 231.5 and Spain's 2,157.8. Greek debt is unsustainable, although in billions it is seven times smaller than the Italian and three and a half times smaller than the Spanish. Crash test evaluations of Italian, German and Spanish banks in late 2016 depicted severe

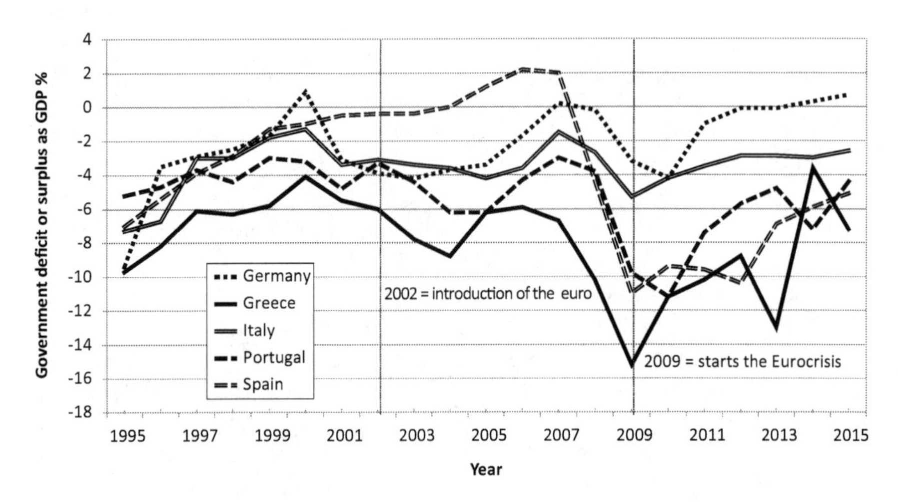

Figure 3.12 Fiscal balance of Germany, Greece, Italy, Portugal and Spain, 1995–2015 (GDP%).

Source: Eurostat, "Government deficit/surplus, debt and associated data", 2016.

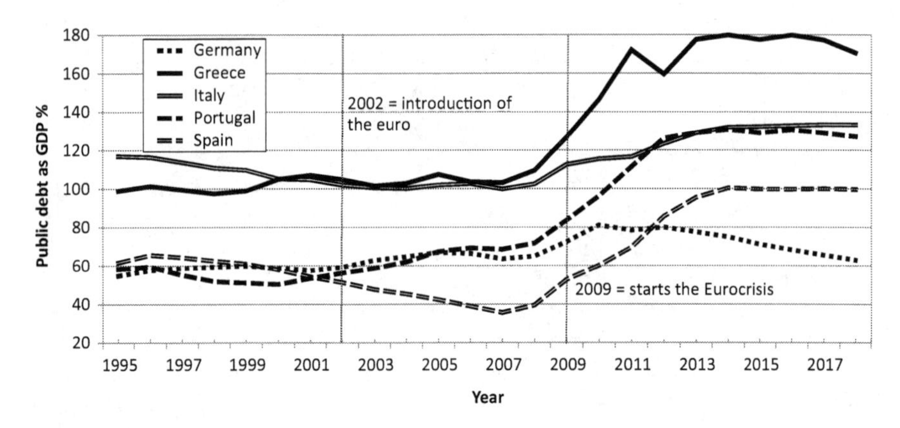

Figure 3.13 Public debt for Germany, Greece, Italy, Portugal and Spain, 1995–2015 (GDP%)*.

Source: Eurostat "Government deficit/surplus, debt and associated data", (2016).

*Public debt refers to government consolidated gross debt.

problems in their balance sheets. Deutsche Bank, one of the largest in Europe and heavily exposed to Southern debts, owns 67 trillion euros in toxic securitisations, what the American financial guru Warren Buffet called "weapons of mass destruction". The Greek debt corresponds only to 0.5 per cent of this exposure. However, any problem in the Greek or in the other SE debts' control may generate domino effects threatening the fragile stability of Deutsche Bank and the entire European banking system – hence, the tough economic, political and ideological control of the South.

If the ECB was a "normal" central bank, public debt may never have become such a problem, so long as it is guaranteed by the central bank that issues the currency in which the debt is denominated. But the European "debt crisis" is mainly a *political problem*, the result of the specific ordoliberal organisation of the Eurozone where, as Fazi (2017: 2) argues, "…governments borrow in what is effectively a foreign currency, i.e. a currency that they don't control". Thus, austerity policies were unnecessary and totally wrong regarding debt reduction. On the contrary, as shown in Figure 3.13, debt has risen precisely as the result of austerity collapsing the GDP.

The creditor–debtor unequal relation in neoliberalism could be seen as a form of value appropriation and circulation over time and space, in which a creditor extracts value from the debtor in the form of interest, i.e. a stable rent-seeking activity. Public debt, as Lazzarato (2012) and Standing (2016) argued, is a brutal political tool to discipline and blackmail populations in SE, first to secure the future of central European banks exposed to Southern debts and second, to make easier fire sales and dispossessions of public property by domestic and foreign capital, supposedly to pay back part of their debt. Greece is, again, a prime example making the link between the politics of public debt and space, particularly public land dispossession (see Box 3.5). It highlights the strategy of capital in the 21st century to acquire profits from rents.

Box 3.5 Public land and public assets dispossessions in Greece during the crisis

From 2010 on, Greece became a target country for public land dispossessions due to the current crisis that has decisively contributed to the de-valorisation/depreciation of land, decreasing monetary values by 15 per cent–30 per cent – depending on the area – when compared to 2005 prices. The special legal status imposed by the Troika forms a lucrative environment for speculators–investors, dramatically altering the legal, constitutional order and imposing something of a semi-protectorate status upon the country.

A key role in the readjusted institutional system is held by the (HRADF) Hellenic Republic Asset Development Fund, known in Greek as TAYPED, inspired by the Treuhand experience, the German agency responsible for privatisation and dispossession of public property in former Eastern Germany. In addition, special planning laws were introduced to make dispossessions attractive to foreign and domestic capital. Because large property is owned mainly by the state/the Church/the monasteries, privatisations and bilateral agreements with global investors take advantage of the authoritarian political framework for quick decisions and clearance sales. HRADF tries to sell public land via privatisations equal to 187 km^2 – a little less than the size

(*Continued*)

of Tinos island – and buildings of 965,732 m^2 – equal to about 52 housing blocks in the Exarchia neighbourhood, Athens' centre. Paradigmatic cases include the former Hellinikon International Airport near Athens; the Piraeus and Thessaloniki ports; 31,000 ha in Halkidiki for gold mining; large public utilities companies and infrastructures that possessed large land plots, such as the (OSE) Greek Railway network, (DEI) the Public Power Company, 14 peripheral airports and many more.

The Greek people, despite strong protest, stand witness to an unprecedented attack against public property for the paying of public debt – which nevertheless continues to increase. What had been won through generations – materially, institutionally and symbolically – is now lost over a small amount of time. The power of lenders and the Troika is so strong that it forced the coalition government led by SYRIZA to continue the implementation of dispossessions decided by previous governments, although this is not an excuse for an illegal and unjust practice that has damaged irreparably its radical left identity.

Source: Hadjimichalis, 2014, 2016.

Eurozone's hybrid space/scale and the undemocratic multi-scalar governance

What is missing from the Eurocrisis debate is the fact that *the Eurozone is a production of a new hybrid uneven space/scale* that is indispensable for the reproduction of European capital and the survival of particular political elites. Space and scale are socially produced rather than predetermined, and the process of EU integration cuts across heterogeneous spaces/scales, linking multiple actors, economic flows and institutions (Swyngedouw 1997, 2000). The new hybrid space/scale of the euro contains several discriminatory geographies, value flows and spatial fixes that reinforce socio-spatial inequalities and hence raise issues of socio-spatial justice, which is discussed in subsequent chapters. Henri Lefebvre argued in the late 1960s that the production of new spaces is one of the solutions available to capitalism when undergoing periodic crises, as a way to absorb surplus capital. He wrote in *The Survival of Capitalism* (1973: 31):

> …Capitalism has found itself able to attenuate (if not resolve) its internal contradictions for a century, and consequently in the hundred years since the writing of Capital, it has succeeded in achieving 'growth'. We cannot calculate at what cost, but we do know the means: by occupying space, by producing a space.

Lefebvre (1976) introduces another important issue in the first volume of *De L' État, L' État dans le monde moderne*, where he discusses scale and argues that "...Today the question of scale inserts itself at the outset – at the foundation, as it were – of the analysis of texts and the interpretation of events" (translated and cited in Brenner, 1997: 137). The "event" of the Eurozone, as the production of new space/scale, is the primary scale of the euro and hides several consequences from us while generating important contradictions and power struggles around the control of particular scales. To again use Lefebvre and his spatial trialectic (spatial practice, representations of space, spaces of representation (see *La Production de l' espace* 1974)), I can see the Eurozone as a new "representation of space" with important economic and political implications. The Eurozone is neither spatial practice (the production and reproduction of particular locations and material characteristics) nor a space of representation (lived spaces, the space of everyday life). It is a representational space only of a currency, the euro, introduced by economic planners and technocrats. During the lengthy negotiations to establish this space, it became evident whose political, economic and spatial interests would prevail, i.e. those of the elites in all countries, but with uneven power (according to country) to implement them.

Currencies in use throughout the world hold no value without underlying sovereign political power at the same scale to make them the legal tender of capitalist exchange. Sovereign states such as the USA, which has the dollar, China the yuan, Japan the yen and so on are able to support their currencies with political and military power. The trouble with the euro is that it attempts to overlay a monetary dynamic *on a political geography and a political space/scale that does not exist as such*: the euro's representation of space is restricted to monetary policies and does not coincide with the sovereign space at the same scale (with a spatial practice and spaces of representation) enjoyed by other international currencies. Writing about this anomaly in 1997, Eric Swyngedouw (1997: 172) argued that the "bumpy history" of the new currency creates an "... uncertain road for the euro...(and) is an example of how a particular and hotly contested politics of scale is inserted into this emerging new scalar gestalt of money".

This *sui generis* situation of the euro, in conditions of crisis, has decisive negative consequences for social reproduction, welfare and labour relations because, as Rodrigues and Reis (2012) argue, "...there is little capacity for managing the tensions in a way that avoids turning labor and social conditions into the main variables of adjustment to crises" (p. 189). Before the crisis, one could argue that this hybrid nature is the euro's novelty and strength. After the crisis, however, it became clear what the Eurozone lacks, which is not only expressed in the so-called "democratic deficit". It lacks the relationship between the sovereign state with sovereign powers and sovereign citizens, not only capitalists but also the people, consisting of all manners of different individuals endowed with the sovereign right to socio-spatial justice.

The operation of ECB exaggerates the democratic deficit in the new Eurozone space/scale. Following its doctrine, financial management should be held by specialised technocrats and not by elected politicians, prone to citizens' demands (Papadopoulou and Sakelaridis, 2012). The social and political control of financial policies is denied, and democracy and accountability are simply cast aside. In this respect, the "political" in the political economy is rejected, and the undemocratic operation is naturalized and de-politicised on the basis of ECB's "independence" and "efficiency". The interruption of liquidity to Greek banks after the announcement of the 2015 referendum is indicative of the ECB's doctrine of "independent" financial policies. At another level, the assumption that in crisis situations technocrats are more responsible to handle financial management, guided the undemocratic replacement of elected PMs by "independent" appointed experts – former bankers, Mario Monti in Italy and Lucas Papademos in Greece.

The emergence of the EU polity challenged state sovereignty, but in a highly differentiated way that depended on the power of individual states. The whole process requires attention to multiple relations between space, scale, governance and power (Clark and Jones, 2008). It represents an example of a new kind of post-sovereign statehood, comprising highly unequal political spaces and scales of engagement (Painter, 2003), what Jessop (2005b) calls a "political meta-governance network". The political rhetoric of European integration silences the intense struggle among the elites to impose an unaccountable scalar governance structure to promote their interests of domination and control.[9]

The first scale concerns labour–capital relations. These are mainly negotiated at national/regional scale, with some minor interventions by the EU, whereas labour struggles take place at local/regional scales. The same is true for social reproduction, whereas inter-capitalist and intra-capitalist struggles cut across scales. The second scale concerns the nation-state where fiscal policy is determined nationally. But this scale is controlled by a third scale – a superior and authoritative one – the Eurozone restrictions of max 3 per cent of GDP national budget deficit and national debt at 60 per cent of GDP. A fourth scale builds around the monetary policy, which is centralised at the EU level with the independent ECB having set the inflation target close to, or below, 2 per cent. Actors at this scale cannot intervene to support individual countries or regions at lower scales when the latter face financial problems. In other words, there can be no bailout, although bailouts were given to save banks exposed to non-performing loans. The state after all is a capitalist state. And a fifth scale is the global scale where financial markets are liberalised and open to all kinds of influence, positive and negative, from events around the globe. These scales are non-hierarchical, constantly re-articulated, highly contested, deeply contradictory and dependent on the outcomes of power struggles.

EU regulations and decisions often become prisoners of this scale structure, which is jumped or reconstructed a-la-carte to adapt to the conjunctural needs of capital accumulation and circulation and the power games of the political elites and capital.

The use of multiple geographical scales also has a neoliberal-monetarist theoretical legitimisation: money and the state should be separated due to the private nature of money, since in the view of ordoliberals, the ECB is independent of governments. This contrasts sharply with the view that monetary sovereignty and the state cannot be separated and that political integration must therefore be followed by monetary integration, not the other way around. Through this contested politics of scale, the capitalist elites of various countries are able to implement policies against labour, social welfare, food regulations and the environment, first at the EU scale, where their decisions are unaccountable, and second, at the national scale by the implementation of austerity measures, arguing that they are coming as EU restrictions. This transnational bourgeois strategy interacts in complex ways with different class interests but with some costs. For instance, the loss of monetary sovereignty has meant that national governments have lost national lenders of last resort. This means that the bourgeois class in a particular country accepts to enter the game in a subordinate position.

Rescaling across and within states under neoliberalism and in the EU finds its application in the systematic construction of elitist policy networks, the marginalisation of already weaker social groups, regions and nations and/or the production of new margins (Hadjimichalis and Sadler, 1995). Following the gradual expansion of the EU, and particularly after the formation of the Eurozone, the European "way of doing things" became less and less democratically accountable due to policy networks that replaced existing forms of government or the building of new ones above or parallel to existing forms. In this multiplicity of scales, only the local and national scales enjoy accountability, with the exception of the federalist systems in Spain and Germany. All other scales are unaccountable. They have their own networks detached from any social control, and this makes the EU and the Eurozone modus operandi authoritarian.

Policy networks discriminate between those nodes that are hooked up to them and those that are not. The tension becomes manifest when modes of justice-based administration and public control that were previously organised on a spatial/territorial basis are replaced by networked forms of governance (Hadjimichalis Hudson, 2004). In these previous modes of local/regional/public administration, citizen-elected representatives and accountability was possible, at least in principle. Neoliberal networked governance is accompanied by the de-democratisation of the political process and leads to political exclusion and limitation of citizenship rights and power (Swyngedouw, 2000).

Unlike the US mode of regulation, the Eurozone and the EU are missing elected bodies similar to the US federal government and US Senate. This absence means that EU and Eurozone institutions lack the legitimacy and policy tools to promote political, social and spatial integration, which, at least in theory, could work equitably and with accountability to help losers from macroeconomic policies. All Eurozone institutions are non-elected bodies. ECB and Eurogroup meetings do not take minutes, and decisions are taken verbally without any transparency.[10] It is not difficult to assume that ECB and Eurogroup presidents could influence policies and decisions in accordance with particular national and corporate interests. This apotheosis of undemocratic decision-making became evident after 2010 during the debt crisis, as exposed by several statements from the former Greek Finance Minister Yanis Varoufakis (Varoufakis, 2016). The multi-scalar and undemocratic governance of the EU/Eurozone is also highly inefficient and slow in decision-making and together with dogmatic ordoliberalism has proved incapable of quick action. It soon became targeted by speculative attacks when the first signs of crisis became visible. In the end, the EU/Eurozone governance processes function along the lines of Gramsci's "ceasarism" and Nicos Poulantzas' "authoritarian statism", with people pushed outside the political arena.

* * *

In sum, the neglect of the uneven socio-spatial structure of the EU and the prioritisation of economic integration at the expense of political integration, together with the preceding financialisation, rent-seeking speculation, real estate bubbles, uneven trade flows and the introduction of the euro, made a volatile mix that with the first external shock exploded as a public debt crisis in the weak links of the Eurozone. Besides the capital switch from the primary to the secondary circuits, I should highlight the inequalities in manufacturing productivity and specialisation existing "behind" the debt crisis. German capital benefitted from falling labour costs, but in any case already was the leader in industrial capital technology, especially after the introduction in 2009 of the Industrie 4.0 programme.

The undemocratic multi-scalar governance of the EU and the Eurozone, plus the functioning of ECB – unreasonable for the people but perfectly reasonable for finance capital – showed their limits with the first crisis sign. Thus, the so-called "debt crisis" is not only economic but also deeply political and geographical. The outcome of interlinked and much deeper structural processes of the Maastricht Treaty, the Eurozone architecture and the geographically uneven capitalist development started as early as the late 1980s and intensified during the 2000s after the euro's introduction, to become the crisis driven forces first clearly seen in the short period from 2009–2010. To solve the problem, another authoritarian network, the Troika, was introduced, this time in partnership with a non-EU institution, the (IMF) International Monetary Fund.[11] It was the last knee-jerk solution by the elites to a development that had not been foreseen.

Notes

1 Among the extensive literature see for Spain: Mart (2013), Soltas (2015), Akin et al. (2014), García (2010), Observatorio Metropolitano (2013); for Portugal: Silva (2013), Romão (2015), Alves Veludo (2014); for Italy: Dinoto (2013), De Cesaris (2009); for Greece: Balabanidis et al. (2013), Tarpagos (2010, IOBE (2015). For comparative studies among two or more countries see: Loureiro de Matos et al. (2015), Siatitsa (2015), Vatavali et al. (2013), Castro et al. (2013).
2 This historical and social characteristic was used after 2011, mainly by German media, to claim that Greeks are "rich in assets and poor in income" and that they should sell their private and public property to pay back their debt.
3 In the USA, 2.8 per cent of the total population moves annually to work in another state, compared to 0.15 per cent in the EU-27 (Eurostat, 2006).
4 The Irish economy, known until the mid-2000s as the "Celtic Tiger", had a similar real estate boom-bust cycle to SE before the 2010 crisis. Its development model, however, differs from SE. It is based on low tax, 12.5 per cent, the lowest in the Eurozone, and on a well-educated English-speaking labour force that attracted large multi-national FDIs in the information technology, banking and insurance sectors.
5 Export figures do not include military and defense weapons, which are subject to specific bilateral agreements. Military spending is a major source of the public debt in Greece, which spends 2.3 per cent of its GDP by 2013. After the USA, the major exporter of weapons to Greece is Germany, followed by France and the Netherlands. See also Chapter 4.
6 Industrial Kombinat was a socialist industrial planning strategy combining Weberian classical locational analysis with technical and economic linkages. The Kombinat was a large organisation usually comprising vertically integrated industrial production activities and social reproduction services.
7 Although the EMU treaty does not allow the public deficit of member states to go higher than 3 per cent of GDP (they are legally subject to a financial sanction mechanism), the Pact was not enforced against Germany and France in 2003 despite their egregious failure to comply with the 3 per cent limit. The European Court of Justice, in July 2004, issued a statement that the Council is not forced to take punitive action against France and Germany, which was a signal for other countries to do the same.
8 Spread is the interest rate differential between two bonds, and it is the speculative way to compare the value of one bond to another. In Europe, is the comparison with German bonds, which are generally considered as the most credit worthy. Bond spreads reflect the relative risks of the two or more bonds being compared. The higher the spread, the higher the risk.
9 I don't underestimate, however, the degree to which scale emerges also as opportunity for popular forces, as an arena for struggle, resistance and solidarity, as discussed in Chapter 6.
10 These practices purposely escape the attention of various EU institutions and international NGOs, such as "Transparency International", which annually

accuses only SE states of a lack of transparency in their bureaucracies. See also Chapter 4.

11 The joint IMF/EU intervention is designed to by-pass several "communitarian restrictions". One of these prohibits the EU from intervening in member states' social and welfare regulation and particularly in labor legislation, health, pensions and education. Hence the joint venture with IMF which bypasses European rules and the so-called "European social model" and could do the "dirty" job.

4 "It is your fault"

Imagining and constructing the new "Southern Question"

The last part of the title is borrowed from Antonio Gramsci's essay "Some aspects of the Southern Question" (*Selections from the Political Writings*, 1926/1978). In his reading of the "Southern Question", as I noted in the introduction, he emphasises the important role of different Italian regional cultural identities resulting initially from the spatially differentiated mode of land ownership and cultivation and later from industrialisation. His analysis provides a de-naturalizing approach to the theorising about cultural/geographical differences. In addition, he paid attention to the "propagandists" of the bourgeoisie, namely the press, the intellectuals, novels and memoirs, promoting a "science" designed to "crush the wretched and exploited", i.e. the different subaltern groups in the South. The ideological operation of dominant cultural institutions, including the Church and language differences, promotes prejudices against Southern people and descriptions of them implying inferiority, which did not escape Gramsci's attention. Indeed, he considers them as the cultural/ideological basis of the "Southern Question".

Drawing from Gramsci, I look on how the new Southern Question – the current crisis in SE– has been ideologically framed by dominant European forces. Different subaltern geographical imaginations of the South were used by Northern politicians and the media to legitimise totalitarian neoliberal austerity. Neoliberalism and austerity are not only economic class projects but more importantly, ideological ones. They try to make us believe that there is no alternative. For example, popular media ask, "You can't do much about the crisis, it's the market, isn't it?" As D. Massey and S. Hall (2010) argued, the market is a powerful ideological and linguistic displacement or transposition of the "capitalist system". It remains anonymous, it erases so many things and since we all use the market every day, it suggests that we are all somehow inside, that we are not only part of the crisis but perhaps responsible for it too. This was the meaning of the statement «μαζί τα φάγαμε» (mazi ta fagame – we ate it together), by the Greek PASOK former Deputy Prime Minister Theodoros Pangalos. He was referring to the wasteful and corrupt spending in the public sector by the Greek political elite, trying to implicate the entire Greek society as co-responsible.

The bourgeoisie and powerful economic-political elites, applying the common "blame the victim" game, use geographical imagination to mask real problems. As Doreen Massey (2012: 29) argues:

> Today, and with unconscious but deep irony, the hegemonic discourses explain the collapse of their geographically inadequate model by turning the blame on to individual constituent spaces (Spain, or Greece) while in fact it is the elites themselves who have produced the problem.

Dominant elites from the beginnings of the crisis, addressing European and national audiences, constructed a narrative that framed as causes of the crisis only Southern Europeans who "live beyond their means" and "their" governments who systematically cheat EU institutions. Unsustainable public debt results exclusively from this irresponsible behaviour, they argue. All other important causes of the crisis, such as the undemocratic structure of the EU and the Eurozone, the greed of the banking and financial sector, the exposure of Northern banks to Southern private and public debt, the unequal intra-European trade, in short the complex and contradictory conditions of uneven development – see previous chapters – are left out of the picture. The narrative of the crisis has regulated what questions should be asked, what experiences count as "best practice" and what principles should be followed. Pierre Bourdieu (1990) calls this "automatization of ordinary logic", an imposition of "unitary thinking" (pensée unitaire). The narrative makes us believe that the crisis comes solely from endogenous factors that made the public debt unsustainable, and it took natural characteristics and framed them with moral and imaginative sentiments. In this narrative, no other system, structure or process contributed to the crisis, only Southern people and their governments are responsible for "their own failure". By the same token, the "victims" of prejudices and negative imaginations often use them to cover their responsibilities as an easy way to escape self-criticism and self-knowledge.

It should be noted, however, that a major difference exists between crises and failures, as Hay (1999) argues. Failure is a precondition of crisis, but not all failures lead to crises. For example, the systematic failure of banks and financial institutions in their predictions about the real estate bubble continued for years until they passed the crucial threshold in 2008. The decades long failure of various EU policies to deliver particular planned outcomes (e.g. the CAP, the structural funds), were never conceived as crises, due to particular "politics of denial" by Brussels bureaucrats. When failures turn to crises, those actors who are capable of making decisive interventions, apart from acting to "solve" the problem, promote a particular discursive construction of the crisis, suitable to their interests. In this mediation, powerful actors played key roles, such as the mainstream media, politicians, bureaucrats and European, national and international institutions, bankers and think tanks. In doing this, old and new imaginations, prejudices and "scientific" analyses were mobilised, targeting mass public moral sentiments and beliefs.

Geographical imaginations

On 1 September 2008, the *(FT) Financial Times* published an article entitled "PIGS in Muck", where PIGS is an acronym for Portugal, Italy, Greece and Spain, starting with the following quote:

> ...Eight years ago, Pigs really did fly. Their economies soared after joining the Eurozone. Interest rates fell to historical lows ... A credit boom followed.... Now the Pigs are falling back to earth.

The article also described that during the credit boom, wages rose and debt levels ballooned, as did house prices and consumption. The use of the acronym PIGS marked the start of an economic and geographical imagery that symbolised the revival of historical cultural and social stereotypes portraying differences between Central-Northern and Southern European societies. The acronym brought back to the surface the colonial power of naming inferiors and provoked an official protest from the Portuguese government. The *FT* was also playing with the English expression "when pigs can fly", referring to something impossible or unreal, as was the success of Southern economies within the Eurozone at the beginning of the 2000s. When PIGS are falling back to earth, they end up in "muck", where pigs belong, making the stereotype even more dirty and repugnant. It reduces Southern societies to subaltern [inferior] humans, "undeserving poor colonial subjects to be reformed or civilized" (Van Vossole, 2014).

The PIGS metaphor (sometimes also PIIGS when Ireland is included) was but one among hundreds used to describe Southern societies and governments. From the beginning of the crisis, an ideological debate raised several moral issues. If SE public debt (plus the Irish and Cyprus cases) is "their own problem", why should other Europeans help? If corrupt governments and irresponsible citizens, such as the Greeks, Portuguese, Spaniards and Italians, have spent "our money" through the structural funds in an irresponsible way, why should we help them again? These issues have been debated for months in the European press (sometimes with vulgarity, see Figure 4.1) and have been discussed in several meetings in Brussels. The ideological debate had a clear dual political scope: first, to promote a negative image of southern countries (governments and citizens alike) as thieves and wasters using old and new stereotypes; and second, using ethical and moral arguments to legitimise the hard ultra-austerity measures imposed by the Troika to reform their "backwards" social, economic and political behaviour.

Figure 4.1 The PIGS eat money from structural funds and EU taxpayers.
Source: www.trumbleart.co.uk.

I do not intend to erase the responsibilities of regions that are lagging behind, and I can offer many examples of EU money being wasted: inadequate leadership, corruption etc. – in SE, particularly in Greek and Italian regions (see next section). But here I would like to discuss the way in which the whole political discourse has been "cleansed" so that class and geographical differences have become "unspeakable" as part of the focus on endogenous factors, of blaming the victim only. In addition, exogenous factors, such as the role of neoliberal pacts in the operation of the Eurozone, unequal trade and uneven geographical development, have also become "unspeakable". Also unspeakable become wider moral issues that are highly related to the present crisis, such as public interest, the common good, socio-spatial justice and their associated principles of democracy and citizenship. Ideology through erasure provides one of the conditions for speaking about, imagining and acting on the crisis in the way the EU/IMF and the media have done.

Edward Said (1979/1996) introduced imaginative geographies in *Orientalism* as the dominating conventional representation of cultures, people and space, as well as of material socio-spatial practices. For Said, containing and representing the Orient through the dominant European culture of the time produced asymmetrical relationships of superiority-inferiority. He adds: "…imaginative geographies not only produce images of the Other but also of the Self" (1979/1996: 77), the "Other" is constructed negatively since he/she is not like Me, B=non A. In a later essay on

Orientalism (Said, 2001/2006), he made clear that his initial book was written not in defense of the Arabs and the Orient, but to highlight different and conflicting "…interests, claims, aspirations and rhetoric that are conditions of open war" (p. 323). This is precisely what has taken place in these years of crisis in Europe. The categories of "Southern" and "Northern" are not empty subjects but full of historical, cultural, political and religious meanings developed under conditions of war, sometimes with military action and these days via financial, political and ideological actions.

As a distinct geographical way of knowing-from-distance, that assigns meaning, behaviour and practices to particular societies and spaces, geographical imaginations have powerful economic and geopolitical effects (Gregory, 1995). They use binary distinctions between "us" and "other" in order to evaluate cultural iconographies negatively or positively. The contemporary racist imagination of PIGS is built on historical and geographical prejudices, which had been lying dormant in the years of EU prosperity, when "PIGS really did fly". For centuries, the imagination and culturalisation of Mediterranean people/places included positive visions such as "perfect holidays", "ancient cultures", "art and music" and "entertainment and enjoyment"; or negatives such as, "liars and thieves", "siesta societies", "sexuality" and "corruption". The current rise of asymmetrical imaginations during the narrative of the crisis builds upon the longue durée of Mediterranean Orientalism. In this respect, it is not the first time that a negative geographical imagination has been used to describe SE societies (see Box 4.1).

Box 4.1 Prejudices from the past

Gina Politi (2016) showed, with examples from 16th to 18th century English literature, the deep historical foundations of negative prejudices regarding Southern societies. In her essay "Demonizing and idealizing the South," "A literary voyage in space and time" (in Greek), she uses the example, among others, of Roger Ascham, who in 1570 published *The Schoolmaster.* Ascham urged young English noblemen to avoid travels to Italy because they would be exposed to vanity, corruption, sexual perversion, Papism and factionalism in politics. The famous quote "An Englishman Italianate is a Devil Incarnate", belongs to Ascham. In the 17th century, Spain became the new demonised protagonist, with which England had various imperialist conflicts. Politi discusses a revenge play by Thomas Kyd, *Spanish Tragedy or Hieronimo is Mad Againe* (1592), demonising Spaniards after the defeat of the Spanish fleet by the English in the 1588 battle.

Later in the 18th century, a reversal took place. Thanks to neoclassicism and the "discovery" of ancient Greco-Roman culture, there arose the idealisation of southern countries as the founders of European civilisation. This idealisation,

(*Continued*)

however, was followed by Orientalism, with a plethora of novels describing Mediterranean societies and places as backwards economically and as sexual Eldorados. Greeks are recognised as descendants of the classical tradition but are not granted full European status due to the long Ottoman rule (Van Vossole, 2014). The same could be said for Southern Italy due to the Spanish and French occupation until the early 19th century and for Andalusia owing to its centuries-long Arab rule.

The dominance of white, bourgeois, Northern-Central, European males, usually blond, over peasant/worker, swarthy, black-haired, garlic-eating Mediterranean people is founded on the appreciation of historical cultural values, while at the same time mounting a colonial cultural crusade to "modernise" them. The duality of demonising/idealising SE continued until the 19th century when, as R. Peet (1985) argues, environmental determinism became "... geography's entry into modern science" (p. 310). He quotes Ellen Churchill Semple, a student of Ratzel in the 1890s, who wrote that for SE and Africa:

> ....northern peoples of Europe are energetic, provident, serious, thoughtful rather than emotional, cautious rather than impulsive. The southerners of the subtropical Mediterranean basin are easy-going, improvident, except under pressing necessity, gay, emotional, imaginative, all qualities which among the negroes of the equatorial belt degenerate into grave racial faults.
>
> (Semple, 1911: 620; quoted by Peet, 1985)

It is worth noting, however, that negative imaginations about the "South" as a metaphor exist also within southern countries for social/cultural groups in particular regions. The classical case is Mezzogiorno in Italy. The imagining about it is in fact based on negation, on what it lacked in relation to the ideal Northern-Central Italy as we showed in the quote from Gramsci. From the late 19th century, dominant descriptions of Mezzogiorno agreed that it was fragmented, disaggregated, corrupt, economically backwards, poor and incapable of self-rule (see among many Nicevero, 1898; Croce, 1925). The South was forgotten even by Christ, as Carlo Levi (1945/1990) famously wrote in his novel, *Cristo si è fermato a Eboli* (Christ stopped at Eboli[1]), describing his exile in Matera, Northern Basilicata:

> But in this dark land, without sin and without redemption, where evil is not moral, but an earthly pain, which is always present inside things, Christ has not descended. Christ stopped at Eboli.
>
> Carlo Levi, p. 4 (my translation)

In Italy, "Italy" means Tuscany, the cornerstone of Dante's language – centre of art and literature. Rome is associated with history and the North means industrial production and hard, stable work. Each of the many different regions and communes in Italy has its own dialect, culture and history. The relative unification of the Italian language became possible only after the spatial diffusion of TV's language in the 1960s, almost 100 years after the end of Risorgimento and the beginning of Rome's status as the capital of the Kingdom of Italy in 1871.[2]

Similar negative imaginations existed in Spain about Andalusians, who supposedly were lazy and took naps daily, the famous siesta, although this was a habit mainly of rich landlords. What people from Castile and Catalonia accused Andalusians of became a general characteristic of all Spaniards in this imagining. On the other hand, Andalusians laughed at Catalans for eating paella every day, even though this dish is a tradition of Valencia and despite the fact that today tourists can find paella all across Spain and in Spanish restaurants abroad.

In Portugal, people from Lisbon and Porto describe the inhabitants of Alentejo, the southernmost rural region of the country, as living a life "at a snail's pace, lazy, mostly old and they are probably communists".[3] Another stereotype about the Southern Portuguese is that they always arrived late to a meeting of any kind, which then became generalised for all Portuguese.

In Greece, the Northern population close to Bulgaria and Turkey, in the region of Thrace – many of whom are Muslims – people are considered inferior and uncivilised by the rest of Greeks and labelled as "Bulgarians" or as "Turks". During the Cold War, large parts of Northern Greece, including Thrace, were declared a "Surveillance Zone", imposed in 1935 and later by NATO as an outpost against communism. After the end of these conditions of exception in 1994, the "Others" in Thrace were "inside" as Greek and European citizens, but remained stigmatised as "outside" and marginal due to religious, linguistic, cultural and spatial factors. There are more similar examples in the four SE countries, but I think it is clear that stereotypes and stigmatisation are not only top-down processes, North versus South, but also work horizontally among members of the same political entity. The reverse is also true, i.e. there are many Southern stereotypes for Northern-Central Europeans.[4]

A crucial dimension in the reproduction of geographical imaginations is economic backwardness. Developmental problems in Mezzogiorno,[5] Thrace, Alentejo and Andalusia were explained solely as an outcome of internal social and cultural factors, local people being "lazy and less energetic", independently of inter-regional relations with the rest of the country and independently of particular interventions by the state and capital. A key starting point is this: "Who frames development problems and how?". After the 1940s and the 1950s, international development theories, inspired in part by the problems of the Italian South, labelled as "South" all those countries and regions not belonging to the "Northern" birthplaces of European and North American capitalism. Theories such as development and underdevelopment, core-periphery, import substitution, strong state intervention in public infrastructure investments and the like came from this period. Economists and sociologists, mainly from abroad and in the case of Italy from Northern Italy, started to use a series of indices such as income and consumption per capita, capital supply, degree of industrialisation, export performance, illiteracy etc. to measure the distance between "underdeveloped" southern regions and countries, with northern regions as the norm. Measuring development distance between nations and regions through such indices became the new dogma and "South" in general became the new international prototype of backwardness. Since then, "South" in economics and development theories has become synonymous with underdevelopment and this has resulted in two major theoretical problems, with devastating effects until today.

The first concerns the indices themselves. Although they appear as neutral, universal and technocratic, they are deeply biased, based on historically and geographically specific social and cultural experiences and choices. Those who use and apply the measures are mainly from "Northern-developed" countries and regions where the historical roots of European industrial capitalism are. They carry as a prototype the particular development trajectory of those areas, which are, of course, different from those places in the South. "Different" here does not necessarily mean lagging, less important or inferior – definitions that derive from the vantage point of dominant formulations imposed by the indices. Measuring development through indices ignores the variety of actual uneven capitalist development, which takes different forms in different socio-spatial formations, as Mingione (1991) has shown in his *Fragmented Societies;* what Bob Jessop (2011) later called "variegated capitalism".

The second theoretical problem concerns the linear, economistic and universal development trajectory assumed by these indices and by those who use them. At one end stand the "developed" countries or regions having the highest or best indices, while at the other end are those "underdeveloped" southern places. The dominant assumption is that "lagging" regions need to "catch-up" with developed ones, and to do this, they need modernisation, outside assistance and a lot of effort. Several decades earlier, the economist Charles Kindleberger called this model the "gap approach": you subtract the indices of underdeveloped regions from the developed ones and the rest is your development programme. In this respect, there is no option for a different development path, no alternatives and the image of the "South" as a metaphor has entered a self-reinforcing cycle in which it is stereotyped.

These issues have been the subjects of further theorising by Doreen Massey in several of her writings. She argues that the way in which space and time are conceptualised in social science (and not only in geography) is of critical importance. Writing about contemporary inequality at the global scale, she argues (Massey, 2006: 95):

> ...inequality that exists within today's form of globalisation...is frequently constructed around notions such as "they are behind", "give us time", "they will catch up". Likewise it is common practice to categorise countries or regions as developed or developing. In all of them, the whole uneven geography of the world is effectively reorganized (imaginatively) in a historical queue...Temporality is reduced to the singular: there is only one historical queue (one model of development) and it is one defined by those "in the lead" (there is one voice) and sometimes, perhaps often, accepted by those who are figured as "behind".

Countries and regions following the catch-up assumption, Massey continues, have no possibility to define a path of their own in space and time: "the future is foretold". In other words, "...this conceptualization of spatial difference as temporal sequence is a way of pronouncing that there is no alternative" (Massey, 2006: 95).

Similar imaginations that turn space into time make it difficult to see, particularly for economists, beyond the present or the Short Past, bowing to "short-termism", as Guldi and Armitage (2014) point out. This dominant vision since the 1980s prevents questions about why and how socio-spatial inequalities established themselves in the longue durée and denies the interplay of exogenous and endogenous factors in reproducing inequalities. These are precisely the dominant views of the Troika's economists, who see only the Short Past of public debt creation, ignoring a larger and more complicated picture of uneven geographical development.

A comparable/complementary distinction to "developed"-"lagging" regions finds application in the significance of language in the production of geographical knowledge.[6] At the present time, dominant geographical discourse originates from Anglophonic literature, which determines the themes and terms of geographical debate. One of the assumptions, determining what "dominant" geographical discourse means is that it produces "theory", issues and concepts assumed to be universally important and relevant (Hadjimichalis and Vaiou, 2004). By this token, theoretical formulations from non-dominant areas and languages are not considered as "theory", but rather as lagging examples, local illustrations or case studies. The latter finds application in the typical division of labour within European scientific teams in which Northerners are responsible in "theorising" while Southerners in doing the "case studies".

The dominant economic and regional development discourse in Europe undeniably originates from languages, places and experiences of central-northern countries that impose a model to exit the crisis defined by those "in the lead", namely the Germans, based on becoming competitive in exports while ignoring severe structural differences among EU regions. Uneven geographical development is reduced to a temporal sequence. This is also the view promoted by most mainstream international media. A notable exception is an article in the *(NYT) New York Times* on

25 April 2010 by Robert Kaplan. Discussing the case of Greece, Robert D. Kaplan, a senior fellow at the Center for New American Security and a national correspondent for the *Atlantic,* entitled his article "For Greece's economy, geography was destiny". Kaplan's article is important because, among other things, it uses geography for the first time in the mainstream media as an explanation of the crisis. His arguments focus mainly on geopolitics and reflect the USA's interests in the region. However, it deserves attention due to his geographical imagination, the colonial language he uses, the arguments about development and underdevelopment and, finally, the use of Fernand Braudel's longue durée as the root of Greece's present problems, ignoring, of course, the disastrous effects of the crisis on local people.

The geographical imagination and historical determinism of Robert Kaplan[7]

Following principles set out in his earlier writings, such as "*The Revenge of Geography*" (2009), Robert Kaplan argues that Greece's budget deficit, its poor economic performance, lack of transparency and abundant corruption all result from its geography. Greece, he says, "…is where the historically underdeveloped worlds of the Mediterranean and the Balkans overlap and this has huge implications for its politics and economy". Kaplan argues that the deeper cause of the Greek crisis is geography, "… (which) no one dares mention because it implies an acceptance of fate". This "acceptance of fate" is a cornerstone of Kaplan's arguments and, at the same time, his primary mistake. Throughout the *NYT* publication, there is a static, deterministic and essentialist view of geography, treating inequalities as an outcome of environmental circumstances. A case in point is his attempt to differentiate Europe's problematic economies in Greece, Italy, Spain and Portugal from the "developed" European core, arguing that all being "in the south is no accident". Although he acknowledges that Athenian democracy and the Roman Republic were innovations in politics, he uses a comment by Fernand Braudel who defined Southern societies as "traditional and rigid". Kaplan argues:

> The relatively poor quality of Mediterranean soils favored large holdings that were, perforce, under the control of the wealthy. This contributed to an inflexible social order, in which the middle classes developed much later than in northern Europe, and which led to economic and political pathologies like statism and autocracy.
>
> (*NYT*, 25 April 2010)

Here, Kaplan makes a direct link with historical/geographical prejudices and echoes Ellen Churchill Semple's environmental determinism (see Box 4.1).

Kaplan's rhetoric reproduces, once more, the dominant European symbolic geography, in which an underdeveloped, poor South is contrasted with a developed, rich, Central-North. This geographical imagination is, for him, an outcome of physical conditions and historical empires, both determining the destiny of places like those in the South. Geographical imagination in turn reflects political opinions in which "undisciplined", "passionate", "cheating" peoples of SE are contrasted with the industrious, rational cultures of the North. And no matter how different the historical contexts, there is a striking continuity in the nature and logic of the rhetoric, as well as in the images and terminology used to represent that dichotomy. These differences are used to define "Europe proper" in terms of inclusion and particularly exclusion, which are central in current political discussions among European elites.

There are several problems with these arguments. First, Kaplan reproduces the old colonial, but still-in-use, stereotype of a problematic South versus a problem-free North and, following a crude environmental determinism, confuses social processes with physical geographical characteristics. A second problem in Kaplan's arguments concerns the relationship between poor quality Mediterranean soils and economic and social development in SE. Here, Kaplan wrongly quotes Fernand Braudel (1966), who among many others is mindful to show that since the 16th century, along with poor soils, there have been historically highly fertile areas, organised not only as *latifundia* but also as *minifundia* (typical SE characteristic), producing for local consumption, but also for export.

Kaplan does not take into account the socio-spatial differentiation of cultivations, which, in a relatively small region, exploit diverse soil types including semi-arid and sodic soils, alluvial irrigated areas, coastal and mountainous terraces, volcanic soils, dry and humid lands etc. Furthermore, he shows his social and political ignorance about a "problematic" SE when he tries to link large holdings with an inflexible social order, statism and autocracy. Southern societies, contrary to Kaplan, developed middle and working classes early on, unlike the dominant Northern industrial model. This was due to the small-holding tenure system and to extensive radicalism among peasant and urban proletarians. With many rural revolts, a widespread anarchist movement, strong unions and strong communist and socialist parties, several civil wars, fascism and dictatorial governments – all during the "short" 20th century as Eric Hobsbawm says – Southern societies could hardly be described as having an "inflexible social order".

Statism and autocracy do exist in SE, yet this is not an outcome of "poor quality large holdings", but rather the last resort of dominant capitalist classes seeking order in a particularly flexible and radical society. And finally, if we suppose for a moment that social and political developments could depend on large holdings, following Kaplan's linear argument, then Northern Europe and the USA – which bypass SE by far in the proportion of large to small holdings – should be the most statist and autarchic areas in the world. Kaplan continues beyond the South-North stereotype to introduce historical-geographical determinism, looking to the schism between Western Europe and Byzantium, the Ottoman capture of Constantinople in 1453 and how mountains divide prosperous areas:

> The Carpathian Mountains... partly reinforced this boundary between Rome and Byzantium, and later between the prosperous Hapsburg Empire in Vienna and the poorer Turkish Empire in Constantinople.... Greece is far more the child of Byzantine and Turkish despotism than of Periclean Athens...
>
> (*NYT*, 25 April 2010)

He is wrongly referring to the "Turkish Empire" instead of the correct "Ottoman" Empire. While he is perhaps correct in noticing that Greek modern culture has more to do with Ottoman tradition than with ancient Greece, it is unfounded to assume that this heritage has in any way contributed to the current economic crisis. Here, Kaplan reproduces a familiar right-wing ideological pillar of the Greek nationalists, who continue to hold the "400-year Ottoman occupation" responsible for all modern Greek pathologies. Blaming history, and the Carpathian Mountains, is a convenient way of evading all contemporary crucial aspects of capitalist uneven development. Kaplan, however, continues and confesses that "...In antiquity Greece was the beneficiary of geography, the antechamber of the Near East...", but a few lines later he argues that "... in today's Europe, Greece finds itself at the wrong, 'orientalized' end of things". In other words, we Greeks are unfortunate to find ourselves at the wrong end of things, inside the "other Europe", which does not belong to Europe "proper"; our destiny is to fail because we are part of the Balkan Orient. This is perhaps the most essentialist and deterministic of all Kaplan's arguments, highlighting his imperialist views, which are developed further in his book *Balkan Ghosts* (1994). By essentialising the "Orient" as "other", Kaplan has assumed to himself that "Western Europe" is the hegemonic partner in the dichotomy. Writing on the Balkans, Kaplan finds recurrent conflict around primordial ethnic identities, and, with the end of communism in Yugoslavia, he hears "phantom voices" that he knew were going to explode once again (Kearns, 2009). Rebecca West (1982), his guide to the awakening Balkans in the early 1990s, argues that the Balkans are popularly defined by violence, incivility and even barbarism, echoing imaginations and prejudices from past centuries (see Box 4.2).

Box 4.2 The "discovery" of the Balkans

Balkan countries were "discovered" as the dark background of the Grand Tour of young noblemen in the 18th and 19th century, who, as part of their high-class education, were visiting Italy and Greece. The focus of their visit was exclusively on the ancient monuments, while all other elements of local culture, particularly those related to Islam, were seen as "...beyond the interest of civilized people", not far from similar, contemporary comments by tourists.

The geographical discovery of the Balkans coincided with the simultaneous inventive/imaginative construction of the area: Travellers' notebooks and geographical books provided descriptions of monuments and the natural environment but very devaluing ones of the local people, since "...they represent a world with less civilized values than the west". The negative connotations of the region and the term "Balkanisation" created a representation of the region in Western thought as troublemaker, to be remembered only in periods of political uncertainty and turmoil. Such images of the Balkans as a place of trouble and violence also permeated popular literature. They inspired Agatha Christie in 1925 to write a detective novel called *The Secret of Chimneys*, in which the story takes place in an imagined small Balkan state where homicide prevails. Descriptions and classifications throughout the novel are revealing of the deep prejudices against the Balkans: the state in question "...is a Balkan country, whose main rivers and mountains remain unknown, although it has many. Capital city Ekaresti. Population mainly bandits. Their hobby: regicides and revolutions." (p. 18). During the war in Kosovo, 74 years later, not surprisingly, similar descriptions were advanced, legitimising the "Balkanisation" of the former Yugoslavia in the 1990s – a state seen as "beyond the pale", not part of civilised Europe. And 85 years later, the Greek debt crisis and the many rallies, often violent, against the Troika and the Greek government refreshed the "troublemaker" description, which was further reinforced by the accusation of handling financial and budgetary issues with lies and trickery.

Sources: Hadjimichalis and Vaiou (2004); Hadjimichalis and Hudson (2003).

As with Semple, Kaplan never uses the West's terminology, although this older symbolic geography of the Balkans as a "dangerous" place is dominant in his book,[8] as is the ideological and political geography of the democratic capitalist West versus the totalitarian, communist East. But in his *NYT* article, he searches for deeper roots to explain today's Europe, using differences between the Prussian and Hapsburg traditions in Poland, the Czech Republic and Hungary, Byzantium and Ottoman traditions:

> ...To see just how much geography and old empires shape today's Europe, look at how former communist eastern Europe has turned out:

> the countries in the north, have performed much better economically than the heirs to Byzantium and Ottoman Turkey: Romania, Bulgaria, Albania and Greece.
>
> (*NYT*, 25 April 2010)

I am not sure that the IMF's bloody intervention in Hungary, the over 20 per cent unemployment rate in Poland and in the Czech Republic (2008) and the degradation of labour and living conditions in these three countries can be considered "better" economic performance. I know, however, how superfluous is an argument, such as Kaplan's, that provides explanations for the current crisis based only on events that took place 200 years ago.

But what will happen finally with Greece with only 11 million people? Does it deserve help from other EU countries? His answer is positive, based on broader geopolitical considerations: it is the only part of the Balkans accessible on several seaboards, has almost the same distance from Brussels and Moscow and is as close to Russia culturally as to Western Europe. Namely, his interest is not to let Greece drift politically eastward, towards Russia:

> ...Germans should realize that Greece...remains the ultimate register of Europe's health. [...] In a century that will likely see a resurgent Russia put pressure on Europe, especially on the former Soviet satellite states in the east, the state of politics in Athens will say much about the success or failure of the European project.
>
> (*NYT*, 25 April 2010)

I know that the decision to "help" Greece and the other SE countries took many parameters into consideration, perhaps including those addressed by Kaplan. But to argue that Germany and the EU must "help" Greece (i.e. the banks that hold Greek bonds) taking into account *mainly* its long coastlines, its equal geometrical distance between Brussels and Moscow, and Greek cultural affinity to Russia (note that for Kaplan Russia is not Europe) is at least exaggerated, not to say simply wrong.

Kaplan's environmental and geographical determinism can be seen as part of a wider political strategy of neoliberal discourse to naturalize socio-spatial relations and conditions and hence to erase the importance of class struggle and all other forms of popular movement. If people in Greece and elsewhere have a destiny to fail because of their (unchangeable for Kaplan) geography, then there is no point in them struggling to make their own future. And as Morrisey (2009) commented about Kaplan's *Revenge of Geography,* which equally applies to his *NYT* article:

> ...perhaps the greater danger in Kaplan's piece is how with feigning intellectual credentials he popularly reinforces a binary that so neatly scripts the necessity of an enduring US military ground presence in the world's key geographical nodes to close the neoliberal gaps of American Empire
>
> (p. 38)

The intervention by the IMF (a direct US presence) and the EU, initially in Greece and later in Portugal, Ireland and Cyprus, is the equivalent of "military ground presence", with the intent to change Greek destiny by an external, enlightened force, this time by closing the neoliberal gap inside troubled Europe.

At work in Kaplan's piece in the *NYT* is a conception of the revenge of geography and history, of space and time, understood in the most old-fashioned colonial sense, which, however, is replicated by many commentators in Europe today. Furthermore, the internationalisation of the Greek crisis and its extension into a Eurozone crisis could be sufficient to prove the inadequacies of Kaplan's arguments. However, I would not like to miss his point about geography, or better about space, not least because many critiques from the left overlook the spatial element built into the crisis and continue to discuss it in a-spatial. macro-economic terms only.

SE countries, including Greece, as Braudel teaches us, developed unevenly along the longue durée, depending on local and global, economic and political conditions. Unlike Kaplan's selective reading of his work, Braudel neither prioritised physical geography nor did he advance any environmental or historical deterministic arguments. On the contrary, Braudel and the French *Annales School* looked at environmental and physical geography as possibilities and constraints, always in relation to human action (including technology) and always paying attention to the world capitalist economy (see also Wallerstein, 1983). For Braudel, economic conditions, space and time were social constructions, with roots in the longue durée but at the same time subjects of ruptures and conjunctural changes from social agents, which may often lead to crises.

If Kaplan were a close reader of Braudel's work, he would recall Chapter II, in Part Two, volume 1 of his *Mediterranean World* on: "Economies: precious metals, money and prices". In this chapter, Braudel describes several economic and fiscal crises in SE, such as the Castilian state bankruptcy under Philip II in 1596 and under Philip III in 1607. An attentive reading of this chapter could help Kaplan avoid his crude link between physical geography and economic crisis. Finally, Braudel never uses prejudices and imaginations for his arguments, as did Kaplan, as well as many others, especially the mainstream media, "the propagandists" of the New Southern Question.

The New Southern Question: prejudices, stereotypes and lies by politicians and the mainstream media

The majority of people need prejudices as digestible and coherent meanings for quick judgment in complicated social situations. We develop stereotypes when we are unable or unwilling to obtain all of the information we would need to make fair judgments about people, places or situations. In the absence of the "total picture," stereotypes in many cases allow us to "fill in the blanks." The mainstream media are a major vehicle for the diffusion of messages containing prejudices and stereotypes among the public. Its language constructs imply public consensus by framing social reality to promote a particular interpretation and/or moral judgment and by intentionally introducing erroneous causes that service the dominant powers. Prejudices and stereotypes are not always wrong and one-sided, but the problem is their partial and often inverted interpretation to produce a particular meaning. In other words, they may be turned into propaganda in the service of particular interests.

In 2010, when the economic crisis hit SE (plus Ireland) a huge media campaign against southern countries took place to prepare the ground for EU/ECB /IMF intervention. They used the classical binary distinction between "us" and "them" and the panoply of traditional stereotypes. Media messages describing Mediterranean people as "lazy" and Northern Europeans as hard-working, were published every week; Southern governments were "irresponsible" while Northerners balanced their spending; and "corruption and inefficiency dominates Southern economies" while virtuous governance prevails in the North; and so forth, the list is long. Political elites reproduce them in official statements in a more politically correct language, retaining, however, the "blaming the victim" strategy. A case in point is Greece, where the mainstream media campaign that accompanied speculation against its sovereign debt presented the country as one that is run by corrupt politicians who "...faked deficit and debt figures to sneak into the EMU" (Schmidt, 2010: 81). A quick search in the popular press, particularly in German mainstream media, is revealing. For example, the front cover of the journal *Focus* (2010) ran the title "frauds in the euro family" with the Aphrodite of Milos' hand giving the finger; and suggestions by the German *Bild* newspaper (27 October 2010) on how to solve the debt problem such as "Sell your islands, even the Acropolis, you bankrupt Greeks"[9] (see Figure 4.2).

I do not intend to deny that in many cases these prejudices contain elements of truth. A case in point is corruption in the four southern countries that is endemic and makes it difficult for ordinary people in the rest of Europe to feel pity for the bankruptcy of the South. Corruption is a popular cliché for Southern Europeans, with Italians taking the lead as "Mafiosi". It is well known that the everyday social and political

Sell your islands, even the Acropolis, you bankrupt Greeks

alle". Aber sie lassen in Deutschland die Schulen verkommen, rauben der Jugend die Chance auf einen Ausbildungsplatz, lassen die Wirtschaft im Stich.

Wir brauchen gute Schulen in Deutschland – nicht nur am Hindukusch.

Über 3000 Inseln hat Griechenland, die meisten in die Ägäis. Nur 87 davon sind bewohnt

Die Regierung in Athen will jetzt kräftig sparen – aber was, wenn das nicht reicht?

Verkauft doch eure Inseln, ihr Pleite-Griechen

... und die Akropolis gleich mit!

Jetzt machen die Griechen ernst, um ihr Land vor dem Bankrott zu retten, ohne auf EU-Hilfe angewiesen zu sein!

Gestern beschloss die Regierung in Athen: Die Mehrwertsteuer rauf (von 19 auf 21%), Alkohol, Luxusgüter und Tabak teurer, die Bezüge von Staatsdienern, Rentnern, Studenten gekürzt.

4,8 Mrd. Euro soll das Sparprogramm bringen – aber bei Staatsschulden von mehr als 300 Mrd. Euro ist das nicht mal ein Tropfen auf den heißen Stein ...

Was kann die Griechen dann noch retten?

Auch wenn es vielleicht verrückt klingt: Wenn wir den Griechen doch noch mit Milliarden Euro aushelfen müssen, sollten sie dafür auch etwas hergeben – z. B. ein paar ihrer wunderschönen Ägäis-Inseln. Motto: Ihr kriegt Kohle. Wir kriegen Korfu.

Tatsächlich ist es der größte Schatz der Griechen: 3054 Inseln, nur 87 davon bewohnt.

Und einen Markt gibt es! Derzeit bietet z. B. das Hamburger Maklerbüro Vladi Private Islands eine unbewohnte griechische Insel an – Verhandlungsbasis: 45 Mio. Euro.

Ob die Kanzlerin morgen mit ihrem Amtskollegen Papandreou bei dessen Berlin-Besuch die Insel-Frage anschneidet ...?

Der Koalitionspartner rät dazu. FDP-Finanzexperte Frank Schäffler zu BILD: **„Die Kanzlerin darf keinen Rechtsbruch begehen, darf Griechenland keine Hilfen versprechen. Der griechische Staat muss sich radikal von Beteiligungen an Firmen trennen und auch Grundbesitz, z. B. unbewohnte Inseln, verkaufen."**

CDU-Mittelstandschef Josef Schlarmann: „Ein Bankrotteur muss alles, was er hat, zu Geld machen – um seine Gläubiger zu bedienen. Griechenland besitzt Gebäude, Firmen und unbewohnte Inseln, die für die Schuldentilgung eingesetzt werden können."

Übrigens: Schon einmal wurde der Kauf einer Trauminsel in Deutschland heiß diskutiert. Das war 1993 – und die Insel hieß Mallorca. Die Idee zweier Bundestagsabgeordneter setzte sich damals leider nicht durch. **Einer der beiden war Peter Ramsauer (CSU) – heute Bundesverkehrsminister ...**

(jan/pro/rok)

Jetzt tickt die Uhr für Athen

Berlin – **Rund einen Monat hat die griechische Regierung Zeit, um mit ihren Reformen neues Vertrauen an den Finanzmärkten zu schaffen!**

Im April muss sich Griechenland über Staatsanleihen 12 Mrd. Euro frisches Geld besorgen. Im Mai noch mal 8 Mrd. Euro.

Hinter den Kulissen ist die Bundesregierung tief besorgt, ob das gelingt – auch wenn Kanzlerin Angela Merkel das Sparpaket der griechischen Regierung gestern lobte: „Ein ganz wichtiges Zeichen."

POST VON WAGNER

Liebe Fußball-Schiedsrichter,

Eure sexuelle Orientierung ist Eure private Angelegenheit. Sie gingen mir allerdings zu weit, wenn ein Fußballspiel gewonnen wird, weil die eine Mannschaft die hübscheren Jungs hat. Der Rasen ist kein Laufsteg. Wir brauchen sachliche Schiedsrichter im Kampf der Dribbler, Wirbler gegen die Rempler, Grätscher.

Das alles wird heute nicht vor dem Landgericht München I verhandelt. Der ehemalige Schiedsrichter-Sprecher Amarell verklagt den DFB, ihm sexuelle Nöti- als Zeuge. „Und er küsste mich."

Wenn mich, Franz-Josef Wagner, ein Mann küsst, und ich will es nicht, dann sage ich: „Noch ein Mal und ich hau Dir in die Fresse!" Meinetwegen können sich die Schiris gegenseitig begrabschen, in Bars, Hotelzimmern. Am Spielfeldrand hat ihre Liebe nichts zu suchen. Ein Fußballspiel ist kein Ballett. Ach, was waren das für schöne Zeiten, als die Schiedsrichter nur blind waren und nicht geil.

Herzlichst

Figure 4.2 *Bild* newspaper article.
Source: www.bild.de/politic/Grienchenland.

behaviour of giving bribes is widespread (*ladono* in Greek, *ungere* in Italian, *untar* in Spanish), in order to navigate the labyrinthine bureaucracy of southern states.

Bild (14 January 2010) emphasises that "...without bribes nothing works in Greece and Italy". The practice of bribing is acknowledged in personal conversations by most Greeks, Italians and Spaniards; but people tend to accuse others, implying that they themselves are not involved. The mainstream media, by criticising local political and social behaviour, the so-called in Greece *fakelaki* (envelopes containing bribes) and *rousfeti* (political favours), look at only one side of the coin. These things exist and happen every day in SE, but first, they are inadequate as *the* economic explanation of the crisis; and second, they refer to the lower level of everyday social practices, excluding large-scale corruption, bribes and political favours among corporate interests, international banks, the state and the EU, which are the real drivers behind the current crisis.

As it does elsewhere, big capital, foreign and domestic participates in large-scale public corruption in southern states to obtain contracts and favours. According to Western mainstream media, these practices are culturally embedded in Southern societies and were used, among others, to legitimise austerity policies. However, large-scale scandals in the supposedly un-corrupt Central and Northern Europe destroy the image that corruption exists only in Southern societies. (VW's) Volkswagen's emissions scandal in 2015[10], the "made in Germany lies", is depicted in a cartoon by P. Chappatte (NYT, 25 September 2015) (see Figure 4.3), where, from a VW car with plates that read "Germany", with a driver and Merkel as passenger, the exhaust produces smoke balls ending in a skull. The cyclist, a Greek with torn clothes, asks (echoing the *Focus* cover) whether Europe will pay the price for German cheating and lying.

VW is not the only big firm involved in scandals; Northern Rock in the UK, ABN AMRO in the Netherlands and Arcandor, Schlecker and Deutsche Bank in Germany remind us that corporate capital is exposed to the culture of corrupt practices at high levels, milking several million euros from it. In Germany, high-level corruption was institutionalised until 2002, and bribes by large firms to acquire contracts, called "useful expenditures", have ceased to be tax deductible. Since September 2002, bribes by German companies to companies abroad are liable to prosecution.[11] Nevertheless, German firms continued corrupt practices abroad, as in the case of Daimler Benz, Ferrostaal and Siemens that admitted to paying tens of millions of

Figure 4.3 "Made in Germany lies".
Credit: Patrick Chappatte/*NYT*.

dollars of bribes to foreign government officials during 1998–2008, including Greece (*Efimerida ton Syntakton*, 10 May 2015, in Greek). In this respect, Berlin's lectures to SE for the last seven years, that the root cause of their economic crisis is their culturally embedded corruption, seem excessive and arrogant.

German and French hypocrisy over the Greek overspending that resulted in the debt crisis is also exposed in the case of the arms trade. Just under 15 per cent of Germany's total arms exports are made to Greece and 10 per cent of France's (*The Guardian*, 21 April 2012). In the period 2010–2011, i.e. after signing the first Memorandum, Greece paid 2 billion euros for German submarines – that then proved to be faulty – which is three times the amount Athens was asked to save through pension cuts. In addition, the arms trade is associated with high-level bribery and corruption, and the German firms Ferrostaal, Siemens and Kraus-Maffei were involved in corruption scandals in Portugal and Greece.

Apart from bribes, a widespread illegal practice concerns tax evasion, a major problem for the state budget. In 2007, the anticipated ratio of tax evasion to total tax in Greece was 28.47 per cent, 26.60 per cent in Italy, 22.40 per cent in Spain and 22.37 per cent in Portugal, while in the Netherlands it was 12.90 per cent and in Germany, 16.37 per cent (Poço, et al., 2015).

Box 4.3 Political and economic scandals in SE

Among the many scandals, here are some examples: In Italy, Tagentopoli, one of the biggest political-economic scandals, began on 17 February 1992 when several socialist party leaders and MPs were arrested for taking bribes. Later, 5,000 public figures fell under suspicion. During 1980 alone, it was estimated that the value of bribes for large government projects reached 4 billion dollars. In Portugal, ex-Prime Minister Jose Socrates (2005–2011 with the socialist party) was sent to jail in 2014 after investigation into his suspected money laundering and tax fraud. In Greece, the ex-minister of defense with PASOK, Akis Tsochatzopoulos, and several of his assistants, went to jail for accepting several million euros as bribes from foreign companies to award them contracts for combat aircrafts and submarines. And again in Greece, several local political officials on Zakynthos Island went to jail in the early 2000s when inspection of the local social benefit budget revealed a high concentration of 600 "blind" people, each receiving a 550-euro monthly benefit while they were healthy and working as taxi drivers, restaurant keepers, farmers, fishermen, etc. The cartoon in the Greek newspaper *Kathimerini* (15 May 2013) by Andreas Petroulakis makes fun of the situation. One person asks: "What is your claim for a union leader's pension?" and the other replies, while reading a newspaper: "I was president of the Zakynthos Blind People's Association".

(*Continued*)

Figure 4.4 Fake blinds: A. Petroulakis cartoon.
Source: *Kathimerini* newspaper, 15 May 2013, by A.Petroulakis.

In 2012, these figures were reduced significantly, but tax evasion remained endemic due to high rates of self-employment and the informal economy. Greek media regularly report cases of tax evasion by specific social groups such as ship owners,[12] doctors, lawyers and engineers but refuse to report scandalous tax avoidance, using loopholes in the law, by ship-owners – the most internationalised part of big Greek capital – and the Church, which is the second largest landowner in Greece after the state. A report in the Greek newspaper *Eleftherotipia* (10 June 2009) about the use of satellite images to discover undeclared swimming pools (which would incur taxes) in wealthy residential districts also made the headlines in German and Dutch newspapers. From a different point of view, and for national political reasons, the UK newspaper *Daily Mail* (29 May 2012) castigated Christine Lagarde, president of the IMF, for her impudence in accusing Greeks of tax evasion while she does not pay a cent of tax on her annual salary of 350,000 euros. It was an indirect response to a previous interview with Lagarde in *The Guardian* (20 May 2012) where she stated that Southern Europeans "had a good time" and "now it's payback time".[13]

The problem, however, is when stereotypical imaginations guide economic policies while data from southern countries shows that the opposite is true. For example, the monotonously repeated argument that Southern people work fewer hours and are lazy is convenient for the elites because it provides an ethical framework for the discussion on the crisis. The South is poor and bankrupt because people there don't work enough; they are "carefree Zorbas". In the Italian daily *Corriera della Serra* (7 September 2012), Massimo Franco published an article on this issue, arguing about

Figure 4.5 Lagarde: "Greeks have to pay!" The Troika refuses a proposition by the Greek government to tax wealthy citizens.
Source: @cococartoons.

the religious roots of this belief. He put forward the hypothesis that "…if the 16th-century German theologist Martin Luther could have been present at Maastricht in 1992, when the foundations of the monetary union were laid, he would have "nixed" the candidacy of the Mediterranean countries". And he concludes by saying that the current bitterness of Northern-Central Europe lies also in the cultural schism between Catholicism, Orthodoxy and the Protestant "work ethic". Josef Joffe – editor of the German newspaper *Die Zeit* – writing in *The New York Times* (12 September 2011), raised a similar point about a "cultural gap" between the "Protestant North" and the "Club Med", pointing to the "happy-go-lucky" Italians. Paraphrasing a popular phrase, he concluded: "It's the culture, stupid!".

The argument about working less in the South, however, is not valid according to the OECD. During 2008, Greeks worked on average 2,106 hours/year, followed by Portuguese at 1,887 hours/year, Italians at 1,807 and Spaniards at 1,713, while Germans worked 1,418 hours/year. Similar data from the OECD regarding average weekly working hours in the EU for 2013 show Greece with 42 hours, Portugal 39.4, Spain 38, France 37.5, Italy 36.9 and Germany 35.3, an average of 37.2 for the EU. Southern people can be accused of being inefficient and less productive, but certainly not lazy. Despite these data, mainstream progressive media, such as *The Guardian* (5 November 2011), continue to use the lazy argument: "…Italian workers have paid themselves more than their German equivalents over the past 10 years for doing less work, less productively". Likewise, it is unjustified and just wrong to seek the roots of the crisis by citing the lack of hard work among the population in SE after so many years of high unemployment. People in SE no longer have the option to work longer hours simply because the crisis destroyed jobs and left millions of people with no work at all.

Another argument influencing austerity policies in SE concerns spending on pensions as a percentage of GDP. For example, EU figures show that Italy, Greece and Portugal have the highest percentages in the EU. However, The *Wall Street Journal* (27 February 2015) took these figures and divided them by the number of people over 65 in each country to come to a different conclusion. Spending per head in these countries is lower than the EU average and behind that of Germany, the UK and France, among others. Nevertheless, politicians insist on using the high pension argument to legitimise harsh pension cuts. In the middle May 2011, Chancellor Merkel said: "…it is important that, in countries like Greece, Spain and Portugal, people should not be able to retire earlier than in Germany" (*Spiegel-on line*, 18 May 2011). According to OECD, however, men in Germany retire on average at 61.5 years, while in Greece it is 61.9, in Portugal 62 and in Spain 62.1. After the Troika's cuts, the average pension in Greece is about 43 per cent of the European average and in 2014 two-thirds of Greek pensioners "survive" on less than 450 euros per month. Obviously, European politicians do not care about the real situation and continue to use prejudice to legitimise their harsh policies. It sounds ironic, however, that Germany is forcing Greece to raise its pension age while it lowers its own. And it is insisting that Greek shops open on Sundays, even though German shops do not.

In the same interview to *Spiegel*, Chancellor Merkel used another cliché for SE. She said: "…We cannot have one currency, when some have a great deal of holiday leave and others very little" (*Spiegel-on line*, 18 May 2011). By blaming Southerners, the Chancellor praises her own people, but a German newspaper argued for the opposite. In July of the same year, *Die Welt* (29 July 2011) had a headline: "Germany is champion of free time". The newspaper cited 2010 research showing that Germans and Danes have on average 30 days holiday leave, above the 25.4 EU average and above the Italians with 26.7, Spaniards 25.3 and Greeks 23. *Die Welt* continues, saying:

> …working people here, including 10 days of public holidays, have 40 days of paid free time, while the EU average is 35.3 days. The usually accused Greek workers and public servants have only 23 days of paid

> leave and 10 days public holidays – this is 17% less than their German colleagues.

And the newspaper continues with a frank comment:

> ...This is a little shock. A self-constructed image is shaken. We imagine Germany as a symbol of diligence. In contrast, we imagine the South, on the sunny Mediterranean, as the home country of dolce vita, siesta and of constant holidays...whereas in poor countries they work more.
>
> (my translation)

Nevertheless, *Die Welt's* statement is an exception. The thrifty North versus the lazy South comes up repeatedly in many media and political statements. Ralph Brinkhaus, senior German conservative Minister, in an interview (*Efimerida Syntakton*, 15 June 2016, in Greek) said that "...hundreds of billions of euros were given to the Greeks. The money did not come from the air. Workers, technicians, clerks have worked hard for them in the rest of Europe, paying their taxes, while Greeks didn't" (see Figure 4.6). This is another myth used to verify the European "solidarity" against Greece, but the data tell a different story.

These billions of euros went to save the banks in Greece, not the Greek people; they also saved banks in Germany, France, the Netherlands and the UK, which were the main holders of Greek state bonds. And they were banking loans to the Greek state that have to be repaid with 4 per cent interest, while German lender banks make huge profits as they can borrow from international markets with 0.4 per cent–1 per cent interest.

Figure 4.6 The myth of European solidarity.
Credit: William Warren/Globe and Mail.

Therefore, by 2015, the Greek crisis and this so-called "solidarity" gave 360–500 million euros of interest profit to the German economy from loans, as Andrej Hunko, Die Linke MP, argued in the German Parliament on 25 April 2015.

These misrepresentations, plus the devastating social effects from austerity, produce hostile public feelings. It is hardly surprising that a sharp rise of anti-German and anti-EU sentiment is widespread in SE and particularly in Greece. Reminders of the German occupation and the many atrocities of the Nazis are vivid among the survivors and their relatives. Chancellor Merkel and Finance Minister Wolfgang Schäuble are the persons most frequently portrayed under titles such a "German occupiers", "imposing barbaric measures" and the like. These anti-German stereotypes resurfaced in the Greek media in parallel with increasing Euro-scepticism. From being one of the most pro-EU populations in the union, Greeks have dropped to one of the lowest levels of support during the crisis. The war between German and Greek media sometimes took extreme dimensions, as a cartoon by T. Anastasiou from *Avgi,* a Greek newspaper, depict W. Schäuble as an SS officer (see Figure 4.7).

We need to remember, however, that the Troika's policies and the turning of southern countries into semi-protectorates could not have been implemented without the active mediation of local political leaders and the local bourgeoisie, who thus found an excellent opportunity to advance their struggle against labour. A case of mediation is depicted in the cartoon by Yannis Ioannou (*Efimerida ton Syntakton*, 27 November 2016, in Greek). Ch. Lagarde and W. Schäuble are sketched as guiding a medieval army, besieging the castle of Greece. A small gate on the left is opened by the Greek Association of Industrialists (ΣΕΒ) and a person

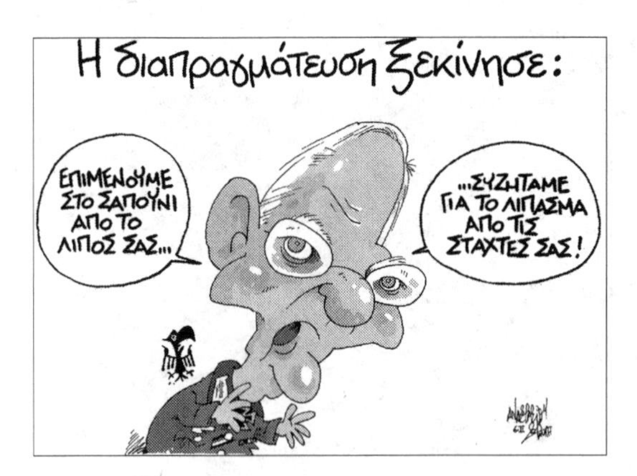

Figure 4.7 Negotiation begins: Schäuble: "We insist on making soap from your fat and to discuss the fertilisers from your ashes".

Source: T. Anastasiou/*Avgi* newspaper.

calls "...Hey!...this way Madame!" – echoing the taking of Constantinople by the Ottomans in 1453, via a small gate, the "kerkoporta" (see Figure 4.8).

Southern elites, capital and politicians, used this long awaited opportunity to implement neoliberal reforms, impossible before the Troika, now shifting the blame onto the "evil of the EU and the IMF". Therefore, it is crucial to remember a few names and political parties active in the period 2009–2015. In Greece, during 2010–2011, the socialist PASOK party was in office, with George Papandreou as PM and later in coalition with the right-wing Nea Democratia with Antonis Samaras as PM. In Spain, the socialist party with José Zapatero was replaced in 2011 by Mariano Rajoy and the right-wing People's party; in Italy Silvio Berlusconi was forced to resign in 2011 and was replaced by Mario Monti, a non-elected banker, decided by the EU elites. Finally, in Portugal the socialist party of José Socrates was replaced in 2011 by the right-wing Passos Coelho. In January 2015, the radical left party Syriza formed a government in Greece in alliance with *Anexartitoi Ellines*, a right-wing party, with Alexis Tsipras as PM, an unwelcome development for the European neoliberal establishment. In July of the same year, however, after a victorious referendum with 61.5 per cent *OXI* (NO) to austerity, Syriza was blackmailed and finally put under immense pressure – some argue for a coup imposed by the Troika – to sign a third austerity memorandum.[14]

A common argument among all SE political leaders, with the exception of Syriza, was that austerity is necessary, replicating the dogma

Figure 4.8 "...Hey!...this way Madame!". Ch. Lagarde and W. Schäuble guide a medieval army, besieging the castle of Greece.

Source: www.efsyn.gr/ioannou. Credit: Y. Ioannou, *Efimerida ton Syntakton*.

advanced by European elites. The same arguments, that local people are lazy, with high pensions, not working enough and always being on strike, were repeated monotonously by national politicians and Southern media, particularly in Spain. The coincidence of bourgeois class interests and neoliberal government practices across national borders, against the interests of local working and middle classes, is no surprise. What is questionable, however, is how the selection of certain endogenous practices, events and groups as the causes of the crisis was unanimously accepted, while those associated with the Eurozone and the EU were put aside by both the EU and local elites. The fact that the crisis caused the debt to rise was turned upside down and constructed with the central claim that rising debt in "profligate" Southern societies caused the crisis. In this causality reversal, of great help was the so-called "Swabian housewife" argument used often by Angela Merkel and introduced in December 2008 after the housing crisis in the USA and the collapse of Lehman Brothers. Addressing her party congress, she used the following quote to explain the turmoil at the other side of the Atlantic:

> What would the Swabian housewife say? "It can't go well in the long run if someone lives beyond their means". And that is the heart of the crisis.[15]

Merkel's statement recycles a naive belief that national economies can learn from these women's frugal housekeeping and operate with balanced budgeting. She, as well as Wolfgang Schäuble, who is also a Swabian, seems to ignore what R. Skidelsky (2013) identifies as the "fallacy of composition": what makes sense for a household or company does not add up to the good of a country. As J. M. Keynes long ago argued, if everyone follows the Swabian housewife's thrift, then total demand will fall, and there will be unemployment for many breadwinners whose housewives will not have much to save. In times of crisis, Keynes continues, this could be disastrous as it contributes to recession.

But Keynes is out of fashion in this era of neoliberalism. So, when the economic crisis crossed the Atlantic to Europe and found new epicentres in the weak economies of SE, the Swabian argument was used both as an economic and ethical doctrine by local and European elites to impose extreme austerity as well as to underpin their ideology by describing people in the South as "wasteful, living beyond their means, preferring lying on beaches". The cynical abstractions and imaginations used by local and foreign elites and the mainstream media build upon pre-existing imaginations and prejudices about subaltern Southern populations while constructing new ones. These are powerful weapons in an undeclared war in which marginal groups, not only in the South but also across Europe and beyond, are the losers. So powerful that after seven years from the beginning of the crisis and the wave of negative prejudices in North-Central press, Jeroen Dijsselbloem, the Dutch appointed President of the

Eurogroup, in an interview to *FAZ* (20 March 2017), made the following comment:

> ...During the crisis of the euro, the countries of the North have shown solidarity with the countries affected by the crisis. As a Social Democrat, I attribute exceptional importance to solidarity. [But] you also have obligations. You can not spend all the money on drinks and women and then ask for help.

His insulting and sexist comment gave rise to heated reactions by all Southern governments, and the Portuguese PM asked for his resignation. But Dijsselbloem declined all critiques and refused to apologise.

The Southern debt crisis, and particularly the Greek, became a powerful symbol, used by speculative mainstream media to instil fear in the European public: "Who is going to be the next Greece?" At the same time, Greece acquired the status of "guinea pig" in the European austerity laboratory and is used to make the "conditions of exceptions" imposed on the country seem like "common sense". The "logical" solution for neoliberals to avoid a Greek tragedy in other countries is to further reduce state welfare expenditure. Hence the crisis in SE acted as a disciplinary mechanism, a Foucauldian bio-political governmentality, helping the emergence of an "austerity consensus" across Europe. Similar structural reforms have been adopted in most European countries, even where there are socialist governments, as in France. In Spain, on 26 August 2011, a few weeks after the end of the occupation of several squares by the 15M Indignados revolt (see Chapter 6), the right-wing government accepted German pressures and introduced into the Spanish Constitution the ordoliberal-balanced budget. The following year, the European Fiscal Compact in March 2012 and the introduction into the German Constitution of a balanced budget, mandatory from 2016, were indicative of the constitutional effects of this consensus, making many speak about a "German Europe".

Notes

1 Eboli is a small city in Campania, Southern Italy, in the province of Salerno. Carlo Levi was referring to Eboli as the location where the road and the railway to Basilicata turn away from the coast, going towards the mountains. See also the film "Christ Stopped at Eboli" by Francesco Rosi (1979) with Gian Maria Volontè as Carlo Levi.

2 See also the famous comment in 1861 by Massimo d' Azeglio, a pioneer of Risorgimento, that "we have made Italy; now we must make the Italians".

3 It is true that ageing prevails in Alentejo, but the accusation of being communists has its roots in the Carnation Revolution, the overthrow of the Salazar dictatorship in 1973 and the nationalisation of latifundia land in Alentejo. Alentejo was a heartland of rural syndicalism and left-wing radicalism and in the 1970s was a strong base of the Portuguese Communist and Socialist Parties. One of the

prerequisites for the accession of Portugal into the EU was the re-privatisation of nationalised land and its return to previous landlords.

4 An interesting collection of stereotype jokes used to compare the South of Spain and in general SE vis-à-vis Northern Europe, is published by Campaña Quién Debe a Quién ("Who is in debt to whom?"), 2011.

5 For a good synoptic description of Mezzogiorno's development since the 1950s, see: Flavia Martinelli (2009) "Casa per il Mezzoriorno", in: *International Encyclopaedia of Human Geography*, London: Elsevier, section on Regional Development.

6 See the special issue in *Social and Cultural Geography*, 2:3, 2001, pp. 261–346.

7 This section draws heavily from Hadjimichalis (2010).

8 The book was highly influential during the Clinton administration and the destruction of the former Yugoslavia.

9 Rosa Luxemburg Foundation (RLF), however, published in 2011 a brochure entitled "Sell your islands you bankrupt Greeks" where our German comrades challenge one-by-one all arguments and prejudices of the German press against Greece and other SE countries (see www.rosalux.de/Greece/prochures). Other institutes and organisations took similar initiatives to RLF across Europe, showing a much-needed solidarity. RLF is part of *Transform!*, a network of 13 European radical and Marxist political and research institutes (see www.transform-network).

10 The VW emissions scandal erupted in September 2015 in the USA, with the carmaker admitting it had installed intentionally software defeat devices in 11m diesel cars worldwide, meaning the vehicles only cut their nitrogen oxide pollution during certification tests. In January 2017, the company itself has agreed to plead guilty to criminal charges and pay a fine of $4.3bn. Six VW executives are charged while in July 2017 other German carmakers are accused for similar scandals.

11 Germany finally signed the UN-convention against corruption only in September 2014, after 11 years of negotiations. See: www.handelsblatt.com/archiv/schmiergeldzahlungen-ins-ausland-nicht-mehr-absetzbar-aufseher-nehmen-exportwirtschaft-ins-visier/2203170.html.

12 In Greece, ship-owners, the most powerful national capital fraction and first in the EU in terms of ship tonnage, paid taxes in 2009 as low as 1.1 million euros, less than the contribution of migrants via applications for work permits (*Eleftherotypia,* 5 May 2010, in Greek). Needless to say, the majority of ship owners do not suffer from the crisis since most of them operate at the global scale.

13 In November 2016, Lagarde was found guilty, but not imprisoned, of her part in a major economic scandal in France, when she was Minister of Economics. Unbelievably, the IMF council decided that there was no reason for her to be deposed from her position as Managing Director.

14 The unjust and forced "agreement" between Syriza and other states in the EU was a defeat for the Greek left and marked the end of an era in which a lot of hopes were invested in Syriza's efforts. Syriza's ministers realised too late that the European elites and the Troika weren't negotiating. They sought compromise and were given fiscal strangulation. After July, Syriza split and the left platform, headed by P. Lafazanis left the party to form a new one, the Popular Unity, favouring Grexit from the Eurozone. In the September 2015 elections, Syriza succeeded in being re-elected into government and since then it has struggled on several fronts at the EU and domestically. According to polls it has lost popular support and legitimacy as the radical left.

15 This simplistic phrase comes from the particular cultural tradition of the Swabia region in Southern Germany, and its message is widely accepted by many in her country. Swabia is part of the rich region of Baden-Württemberg and famous as a caricature of middle-class ideology but also for hard-working people who carried out bloody savings and prudent management. The German original text is: "Was hätte die schwäbische Hausfrau denn gesagt? Es geht nicht gut, wenn man auf die Dauer über seine Verhältnisse lebt. Und das ist der Kern der Krise".

5 De-politicising uneven development and socio-spatial justice

A key component in the implementation of neoliberal policies is to frame development problems as repairable by technocrats so that development policies are formulated in an a-political and often a-spatial framework. In this sense, development decisions are in line with Marx' idea of de-politicising issues whereby they are removed from the level of public accountability and designated as non-political. In this respect, they block the possibility of debate and struggle within society. Academic discourses since the 1990s have contributed, intentionally or unintentionally, to the reproduction and legitimacy of the shift towards de-politicisation, first by naturalizing or essentialising concepts and second by finding ways to side-step politics, the principal one being to avoid identifying winners and losers. It should be noted, however, that the neoliberal fetishism of an a-political market is highly political indeed. It serves the interests of the bourgeois class, particularly its financial and banking fractions, and works to the benefit of power elites against representative governments. The supposed "neutrality" of de-politicised markets is nothing less than the political power game of capitalism.

A necessary condition for the implementation of neoliberal policies is the distortion of language. Neoliberalism has its Newspeak and an army of communication specialists whose mission is to deform reality (Massey, 2015). De-politicisation imposes a set of unspoken rules, which silently but powerfully determine what can and what cannot be said. Vague technocratic descriptions try to cover bloody attacks on social welfare, marginal peoples and regions. In this spirit, when government officials and technocrats speak about modernisation and structural adjustments, they mean budgetary cuts for health, education, social services and the dismissal of public servants. If a working person becomes redundant, they lose their job and consequently their social security. It is described as an increase in flexibility and efficiency of social security. And when they impose crude austerity on whole societies, they speak simply of structural reforms to increase competitiveness. Therefore, it should not be surprising that a culture of fear arises, alongside feelings of injustice and anger, together with sub-cultures proposing alternative political solutions of hope, albeit not always progressive.

Although the economic crisis deepened the crisis of political legitimacy, aggressive individualism and collective punishment – among the cornerstones of neoliberal Newspeak – are dominant. Hence, the leading German newspaper, *(FAZ) Frankfurter Allgemeine Zeitung*, uses the mantra "without pain (austerity) there will be no gain (structural reforms leading to growth)", a prime example of Newspeak that makes perfect neoliberal sense, as in some cartoons that display Spain's flag after "structural reforms" without the "S", reading "PAIN". Other important voices in Europe, however, from progressive and radical individuals or from political parties, stand up and attack the essence of neoliberal language and practice. Among those, the President of the Belgian Magistrates' Union, Manuela Cadelli (2016), vocally argues, "neoliberalism is a species of fascism because the economy is brought under subjection...every aspect of our thought and speech", and she describes how deep the principle of budgetary orthodoxy wend, to be included in state constitutions. She continues:

> The austerity that is demanded by the financial milieu has become a supreme value, replacing politics. Saving money precludes pursuing any other public objective. [...] The nihilism that results from this makes possible the dismissal of universalism and the more evident humanistic values: solidarity, fraternity, integration and respect for all and for difference
>
> (*Le Soir*, 3 March 2016)

These voices, however, are not enough to reverse the material deprivation resulting from the crude austerity policies pursued before and during the crisis across Europe and particularly in the South and East. Socio-spatial inequalities increase at every spatial scale and in every aspect of everyday life, affecting directly the "bio-politics of population," as Michel Foucault would say. The neoliberal attack against the weak welfare state in the four southern countries imposes a "regime of precarity" (Athanasiou, 2012), regulating poverty and life itself. Health, social reproduction, sexuality, life expectancy and so forth become part of "precarious life", the neoliberal strategy to control and govern. These developments were visible long before the current crisis unevenly affected individual states, regions and cities, but they took extreme dimensions after the imposition of the painful and unjust "structural adjustments" as a kind of "collective punishment" for the debt failure, following the *FAZ* mantra. With cruel irony, during a period when poverty, income inequalities and socio-spatial polarisation in Europe were reaching unprecedented levels, local/regional welfare planning almost disappeared. The dominant model became the entrepreneurial city and region.

Paradigm shifts in economic geography and regional development

In the previous chapters, I argued that the crisis in SE is a symptom of deeper causes related to financialisation and to a variety of endogenous/exogenous aspects of uneven development, in which the neglect of geography and space became a key component of the crisis. If my argument about the social and geographical/regional foundations of the current crisis in the EU and the Eurozone makes sense, what was the reaction, vis-à-vis the causes and the effects of the crisis, from senior researchers in the fields of economic geography and regional and urban planning, which are the primary disciplines handling these issues? Apart from the "usual suspects", such as D. Massey, D. Harvey, R. Hudson, E. Swyngedouw, F. Indovina, E. Mingione, M. Garcia, D. Vaiou and a few others, to my knowledge at the time of writing, it was very limited indeed – almost a complete silence.[1]

This is nothing to be surprised about, as the dominant policy views on European integration over the last three decades have been neoliberal, and when they speak of geography and regional development, they don't have much to say that is critical. We need to remember that neoliberalism in Europe and globally is far from homogeneous; there are different national and even regional strands, which lead us to talk about "really existing neoliberalisms" (Peck et al., 2010). Or to put it in another way, we have uneven impacts of the neoliberal doctrine on different national/regional social formations. With important differences among states and regions, the neoliberal turn in economic geography and regional and urban affairs has neither raised cities and regions to new levels of dynamism (apart from a few superstar cases), nor diminished capitalism's inherent tendency towards crises and uneven development. It has created a major market for new policy "discourses", which have been met – thanks to the parallel neoliberalisation of academia – by the creation of new degrees and professional courses (Lovering, 1999).

On the theoretical front, the financial crisis in SE, Cyprus and Ireland came at a time when economic geography and regional development in Europe was dominated by two major paradigms, both in use by academics and practitioners alike in member states and by the Commission. The first is straightforwardly neoliberal, based on a modernised version of old neoclassical theories and macro-economic top-down planning, loyal to the dogma that free market competition will eventually balance inequalities. It has excessive faith in highly abstract mathematical models as the basis of regional science and policy practice, assuming that poorer regions tend to have higher rates of growth than richer ones and that convergence is a matter of time (Barro and Sala-i-Martin, 1995). Spatial modelling went hand-in-hand with financial and real estate modelling, both based on the prediction that prosperity is secured through rent extraction. Shut hiding behind the assurance of prosperity during the early years of the strong euro, the neoliberal paradigm has succeeded in directly influencing the Commission and its technocrats.

It is hardly surprising that the first paradigm had little to say that was self-critical about the crisis in Europe. What seems more surprising, though, is the silence of those progressive and leftist senior colleagues who, following a kind of "Third Way thinking" – the second dominant paradigm – were responsible for developing new theoretical approaches shaping local and regional development policies before and after the crisis (Hadjimichalis and Hudson, 2014). Colleagues working within this paradigm, parallel to more traditional approaches of agglomeration economies, transaction costs and external economies, promoted new ideas on innovative, networking and learning regions, on clusters and agglomeration, on branding, on local social capital, reciprocity, trust and so on, largely based on experience of industrial districts in Third Italy – people who today remain silent. Key advocates include Cooke and Morgan (1998), Amin (2003), Asheim (1996), Becattini et al. (2003), Storper (1997) and Scott and Storper (1988), among others.[2] Many other economic geographers and planners made a strong appearance in EU circles when they started working within this second dominant paradigm, which also had a major influence on EU development strategy.

As is by now well known – and so just briefly mentioned here – this "Third Way thinking" is identified by two major schools of thought: first the so-called "(NEG) New Economic Geography" or "geographical economics"[3] and second, *(NR) New Regionalism.* Without forming a coherent body of theory, NR promotes several proposals for "learning, networked and flexible" regions, "innovative and intelligent" regions, "clusters", "creative" cities, "communicative planning" and the like. They appeared within the panoply of progressive local/regional development theories in the 1990s (Asheim, 1996; Cooke and Morgan 1998; Healey, 1997; Storper, 1997; Morgan, 1997; Amin and Thrift, 2005).

In Europe, "Third Way thinking" and particularly its New Regionalism branch became highly influential among EU policy makers, particularly in the crucial period 1994–1999, during which the Eurozone preparations took place (Thoidou and Foutakis, 2006: 31–32). From the promotion of infrastructures to allocating funds to lagging regions and handling unemployment, spatial cohesion policies shifted to entrepreneurship, vocational training and institutional restructuring. Technological innovation and human capital were the two pillars of the new policy following the new EU Development Strategy, introduced in 2000 with the Lisbon Pact. Although initially a social-democratic project, it was taken over by the EU technocrats and has been transformed into a neoliberal project. The new strategy establishes national and regional competitiveness, innovation and the "learning economy" as the main EU planning targets.

Thus, "Third Way thinking" became the new orthodoxy in progressive economic geography and regional studies and influential in local and regional development policies in Europe and beyond. Of course, one cannot blame original ideas if others use them wrongly. All I am saying is that the initial selective appropriation of Third Italy's history and the downplaying of other important aspects of this particular developmental path opened the door for many one-sided or mistaken interpretations regarding the wider SE development path, and this in turn was a major handicap/barrier to understanding the dramatic changes in the 1990s and 2000s.

Take for example the emphasis on non-economic factors in regional/national development that is a major pillar of "Third Way" thinking and highly influential in EU policies. The origins of these arguments can be traced to Robert Putnam's work on Italy (1993) that conceptualises non-economic characteristics under the term *social capital*, and later to Porter's work on clusters (1998).[4] For Putnam, the agent is represented through trust and reciprocity as a socially individualistic, abstract masculinity, which is somehow magically transformed into peaceful and collaborative communities, regions or even nations. The social actor is inherently male, and in this sense, social capital theorists differ very little, if at all, from the theoretical assumptions of rational choice theorists where the actor is omnipotent without formal or informal ties to others unless they are contract related.

Box 5.1 Critique of social capital

Social capital in Putman's work and its influence on the "Third Way" literature is highly questionable. Attaching the adjective "social" to the economic term "capital" suggests, at least in the "Third Way" approaches, that along with non-social forms of capital (such as finance capital, material capital, industrial capital, etc.), there exists a social variant of capital. Conversely, the modification of "capital" as "social" signals the attempt to overcome the difficulties of dominant neoliberal economic models and thus draw attention to neglected non-market conditions of economic growth and social development. The major problem with this formulation is that it reproduces a 19th-century view of capital only as a "thing" and confuses capital with money or with other material forms such as machines, raw materials, energy inputs, products etc. Marx was critical of these approaches and in *Capital* vol.1 defined capital both as a material form and a *social process* rather than solely as a thing. In capitalist societies, this presupposes the existence of class relations between capital and labour. Thus, the adjective "social" is a mere confusion, since capital is anyway social. This is perhaps the reason why Pierre Bourdieu is never mentioned in these studies, although he was the first who worked on these issues and his work is also translated into English.

Bourdieu's (1980) main conceptualisation of social capital refers to individuals and rests upon the query of how cultural reproduction fosters the social reproduction of relations between groups and social classes. He defines social capital as "...the aggregate of the actual or potential resources which are linked to possession of a durable network of more or less institutionalized relationships of mutual acquaintance and recognition" (1980: 248). Putnam defines it as: "...trust, norms and networks that facilitate cooperation for mutual benefit" (1993: 167). He refers to communities, regions or nations (hence its attractiveness to geographers and regional scientists), and he measures it in terms of participation in voluntary associations. Here we have two opposing sociological traditions, that of integration (Putnam) and of struggle (Bourdieu). For Putnam and his followers, social capital manifests itself through a process of friendly negotiations for the benefit of the entire community. For Bourdieu, social capital mostly develops through the economic infrastructure and is constituted by individuals struggling for power and recognition, not mere cooperation. While for Putnam and co. it's invariably benign, Bourdieu's concept invites a more critical view.

I believe that the treatment of social capital, trust and reciprocity in "Third Way" studies and policies, rather than enabling us to reconceptualise non-economic factors in terms of remembering the political involvement of human agency (supposedly forgotten in *all* Marxist studies), takes us back to the instrumental management of resources and makes development at various scales a matter of voluntary civic engagement. This does not take into account the inherent conflicting parameters characterising politics. This way of building the social into development appears to be a powerful means of justifying economic neoliberalism. The main assumption, central to this conceptualisation, is that social interactions are conflict free and can mobilise resources for the benefit of the entire community. In this respect, the context is de-politicised, making us free from politics but not from economics.

One-sided analyses of non-economic factors are but some among the many misinterpretations of "star" regions. The most important problem came later when something ominous began to happen in Europe in the early years of the euro's introduction. Just at the very moment that policy prescriptions following the "Third Way" paradigm, based upon the assumed bases of success of Italian and other localities, were being generalised through the design of regional and local development policies in the EU, the conditions on which success was based in these exemplar regions were being eroded (see Chapter 2). In this respect, when the crisis began in 2009, both dominant paradigms in economic geography and regional studies, first the modernised version of old neoclassical theories and second the "Third Way", were inadequate to understanding the spatial component of the crises and were caught on the hop. They remain unable to incorporate its multi-scalar geographical/regional causes.

Despite its progressive intentions, "Third Way" treatment of urban/regional development issues is often highly compatible with a neoliberal view. I don't argue that theories of NEG or the NR models are neoliberal *stricto sensu*, or that their exponents are neoliberals. Neither is my aim here to take issue with colleagues with whom I agree on many other counts. All I am saying is that in the crucial decade of the 1990s, the way they formulated, directly or indirectly, the original question posed by D. Massey back in the late 1970s, "In what sense is it a regional problem?" de-politicised it, at the same time as neoliberalism was making a frontal attack in the field. Thus, the highly needed resistance and fighting back never materialised from "Third Way" researchers. This has made their views easy to absorb into neoliberal policies, making it sometimes difficult to differentiate progressive from regressive applications. In that sense, there was a gradual sliding towards the dominant neoliberal discourse. Let me clarify this with five points:

1 A key difference over values, goals and methods between "old" welfare and "new" regionalism is political: In the former, welfare distributional questions (between people and places) directly connected regional policy with politics. The latter treats regions and cities as quasi-individuals, obliged to find their own ways to economic prosperity, competing with others (Hadjimichalis, 2006; Lovering, 1999). What regions (or cities) need, according to "Third Way", is less politics, more competition, more innovation, more pluralism, learning and tolerance. In this respect, "Third Way" has been associated at the sub-national scale with the wider de-politicising, which has been a key objective of neoliberalism at the national and global scales.
2 NEG and NR theories paid attention to a few successful superstar regions and cities (Perrons, 2004), neglecting all other "ordinary" places under the assumption that superstar success came primarily from internal, endogenous factors within the region or urban area in question while ignoring exogenous forces (Hadjimichalis and Hudson, 2007, 2014). These assumptions were based on a cognitive shift towards seeing each place as a bounded entity to be studied in its own right. This is highly compatible with the neoliberal discourse, which promotes the success of the few, idealised, competitive individual efforts, while ignoring relational politics and downplaying wider social and spatial conditions, particularly those in "failed behind" areas. In addition, this particular focus of NEG and NR approaches helped the establishment of arguments that the causes of the crisis were and are only endogenous (e.g. corrupt governments, cheating citizens etc.), ignoring exogenous forces such as the very modus operandi of the Eurozone.
3 Following the previous emphasis on regions and urban areas as the pivotal spatial scale for capitalist success, these approaches ignored the regulatory role of the state and particularly its potential in the struggle to ameliorate the lives of people in these places that "failed" (McLeod, 2001; Hudson, R., 2007). In a period in which major re-scaling of governance and the widespread introduction of public-private partnerships took place across Europe, NR failed to see the sliding towards de-politicised governance and continued to focus only on particular

successful regions and cities. On the one hand, it underscored the role of the state as again compatible with the neoliberal dogma "less state – more market", followed by class-based policies for deregulation of the welfare part of the state and massive reregulation to meet the needs of capital. On the other hand, the over-emphasis on regions has not really helped these approaches to understand the geographical/regional foundation of the current crisis and to realise the major multi-scalar governance change introduced by the Eurozone.

4 Although these mainstream views pay attention to particular regions and cities, and have provided pioneering analyses of local productive and institutional structures, they over-emphasise the supply side, giving scant, if any, attention to understanding the empirical dynamic of the demand side and of global capitalist competition in the sphere of circulation. Within the Eurozone, the question of employee compensation is crucial and became a major competitive element after the introduction of the euro. The inadequate analysis of the commercialisation, distribution and retailing of products and services coming from model regions has become a major handicap, realised only after 2000 and with the first signs of crisis in 2009. The emphasis on the supply side helped to mask unequal trade among Eurozone regions, in particular how debts in the European periphery are related with trade surpluses in the Central-North.

5 The lack of interest in social issues beyond success and competitiveness is also evident in inadequate analyses by NEG and NR of issues of everyday life and living (beyond consumption and lifestyle) including work, poverty and unemployment and the current deep social, demographic and ethnic changes in Europe at multiple spatial scales. Focusing only on technical and procedural modes of flexibility and innovation, they play down the role of labour, and they ignore class, gender and ethnicity, (Perrons, 2012a,b; Karamessini, 2013) thus losing contact with everyday life in ordinary firms, regions and cities. This is also evident in how the de-politicised new language of EU/IMF intervention in SE, such as "helping", was received by the "Third Way" politicians of southern countries who have used the same language to try to convince the public that cuts in social services and wages are "necessary" as "medicines" for the debt "disease".

These five points summarise the distance of "Third Way" ideology from older progressive formulations of welfare regionalism, not to speak of radical leftist views of uneven regional and geographical development. However, NEG and NR cannot be blamed either for all the problems in the Eurozone, or for EU/IMF intervention. All I am saying is that the dominant discourse in economic geography and regional development by sliding, consciously or unconsciously, towards neoliberalism has proved itself unable to understand the geographical/regional foundation of the crisis and has helped to direct regional development questions towards non-combative paths by de-politicising them. While such omissions would be expected from a neoliberal perspective, it is rather puzzling for me that radical senior theorists and researchers succumb to the charms of grand narratives, even when they strongly argue for the need to pay attention to differences and to local processes.

Perhaps the largest failure of both neoclassical and NR approaches is their neglect of periodic capitalist crises, what neoliberal economic advisors call "systemic failures". This negation is typical of all apologetic analyses of capitalism and something to be expected from the neoliberal side. But it is surprising that very few "Third Way" researchers have paid attention to the capitalist crisis already visible from the late 1990s in Third Italy and other emblematic regions. The trouble in these times is that most people have no idea who Keynes was and what he really stood for and do not know Myrdal's "cumulative causation" and "backwash effects"; uneven development sounds "too political" while the understanding of Marx is negligible.

When the economic crisis began in 2009, at a time when NEG and NR were highly appreciated by technocrats and development officials, there was no other paradigm except the neoliberal one at the table of the top EU, Eurozone and ECB bureaucrats and state politicians. Different economic advisors went deeper, suggesting that austerity *is the solution*, and while initially it may cause pain and loss (see the mantra of the *FAZ*), in the long run it promotes growth. Today, such an argument sounds ridiculous, but at that time, it generated some disagreements among EU top officials, between "austerians" – those advocates of austerity and cuts in public spending – and a few neo-Keynesians who believed austerity functions only when the economy is growing. Among the "austerians", two prominent Italian macro-economists, Alberto Alesina and Silvia Ardagna (Alesina and Ardagna, 2009) wrote a series of influential papers attacking the neo-Keynesian arguments that cutting spending during crisis in weak economies makes the situation worse. Alesina and Ardagna provided "scientific" support to neoliberal dogmatism and for this were invited in April 2010 for a special presentation by the Economic and Financial Affairs Council of the European Council of Ministers. Their analysis was highly appreciated by the European Commission and the ECB. Thus, in June 2010, Jean-Claude Trichet (*Reuters*, 24 June 2010), the then President of the ECB, dismissed concerns that austerity might hurt growth:

> ...As regards the economy, the idea that austerity measures could trigger stagnation is incorrect.... In fact, in these circumstances, everything that helps to increase the confidence of households, firms and investors in the sustainability of public finances is good for the consolidation of growth and job creation. I firmly believe that in the current circumstances confidence-inspiring policies will foster and not hamper economic recovery, because confidence is the key factor today.

Thus, by the summer of 2010, a full-fledged austerity orthodoxy took shape, becoming dominant in European policy circles to the extent that the formation of the Troika and the implementation of crude austerity in SE became naturalized, ignoring all other "softer" proposals by neo-Keynesians and "Third Way" policy makers.

Socio-spatial justice after the financial crisis

Although uneven geographical development is associated with socio-spatial inequalities, not all social processes take place uniformly across space, because geography, by its nature, is always uneven. Socio-spatial inequalities arise from and are reproduced by the uneven geography of capital accumulation and flows of capital, from uneven spatial divisions of labour, from actions by social groups and from particular state policies and other interventions. In the current conjuncture, the particular neoliberal accumulation and governance regime represents a radical de-politicisation of structural inequality, along with an increased tolerance for it. Socio-spatial inequalities are framed not as structural problems but as the result of wrong choices or failures individuals have made, without any regard to any factors that might have constrained those choices or failures. Thus, the decline of critical and radical currents of thought in economic geography and local/regional development – within "Third Way" thinking – leads to an impasse within academia and among policy makers, perhaps comparable to asking the bankers who created the current financial mess to clean it up with exactly the same tools they used to get us into it. Ray Hudson (2007) summarises the issue:

> ...Uneven development is an integral component of capitalist economies and while some regions will exceed national (or other) growth rates and targets, others will not. In other words some will "fail" as part of the price of others "succeeding"....The mainstream view pays scant, if any, attention to distributional issues as central to development and there are strong grounds for arguing that issues of social and environmental justice and equity must be central to any sustainable economic development strategy.
>
> (p. 1156)

The above discussion by Hudson directly introduces the problem of the values and goals of regional development in an unequal Europe; in other words, the politics of progressive planning. Dealing in space with losers, and not only with winners (as "Third Way" does), re-introduces the notion of politics to thinking about space politically and politics spatially, hence recognising a responsibility for socio-spatial justice. Winners and losers are not social and spatial aggregates, but firms, social classes and individuals living and operating in particular places and regions, which gain or lose due to the particular unequal economic and power relations in which they engaged. Socio-spatial inequalities derive from the inevitable outcome of the uneven geographical development in capitalism. Socio-spatial injustices are the (bodily) lived and experienced reality of those inequalities in particular places at different scales – the "lived spaces of our daily lives", as Henri Lefebvre argued long ago.

Doreen Massey (1999) uses a particular expression for this. She calls it "the geometries of power", that is the complex relations of domination/subordination and marginalisation/exclusion through which citizens, firms, regions and states engage in everyday life and in politics, which inevitably leads to socio-spatial injustices. She goes on to call for "geographies of responsibility" (Massey, 2004, 2007) among people, places, cities and countries with different levels of power and wealth and different identities and histories. In the absence of such a discussion, the majority of analysts – including some from the left – remain within the mainstream policy environment, which favours interventions targeted towards either reducing the costs of doing business or improving the competitiveness of firms, cities and regions. Such emphasis ensures that theory is invoked to justify current practice, further diverting attention from the deeper causes of local/regional deprivation (Pike, 2007; Pike et al., 2007).

Daniel Dorling (2010), in his book *Injustice: Why Social Inequality Persists* introduces five key tenets of injustice: Elitism is efficient, exclusion is necessary, prejudice is natural, greed is good and despair is inevitable. Although these tenets have very old origins, nevertheless today they make perfect neoliberal sense and seem particularly relevant to interpreting the present conjuncture. Dorling makes a direct correlation between injustice and social inequality, the first deriving from the second. Although he does not refer to space, the spatialisation of his five tenets seems obvious. We can perceive the dogma "elitism is efficient" in policies promoting a few super-model regions and cities as "intelligent regions" or as "creative cities", ignoring the majority of ordinary places. Gentrification projects, reducing funds to poorer regions and restructuring social housing in cities, make "exclusion necessary" to the benefit of capital, although the regional technocrats and landlords argue for the opposite. The "naturalization of prejudices" finds application in geographical imaginations of "the Other", native to peripheral areas, as subaltern. "Greed", the unspoken principle in all business and management schools, "is good" when it mobilises capital's motivation to de-localise factories and production, banks to accumulate super-profits through the securitisation of housing mortgages and rich individuals to become billionaires, acquiring property from fire sales and from dispossessions of public land and former public utility companies and turning them into assets providing rents. And of course, "despair is inevitable" in foreclosures when localised unemployment increases or when people lose their pensions and social security, as in SE. We can read these principles of injustice behind all of the Troika's arguments, and it is no secret that those who believe in them belong to the dominant classes.

Moving now to spatial justice, David Harvey (1973) in his *Social Justice and the City* came close to the concept with the use of territorial justice as the just distribution of social resources and goes on to define the processes that produce spatial injustice. He argues, "...space, social justice and urbanism cannot be understood in isolation from each other" (p. 17), and for Harvey, social justice is not a matter of eternal justice but as "...something contingent upon social processes operating in society as a whole" (p. 15). Just distribution or distributional justice means to reduce social inequalities and to bring the greatest benefits to the least advanced. In a later essay (Harvey, 2009), he argues that any attempt to achieve a just city within the bounds of capitalism is doomed to failure. In their edited volume, *The Urbanization of Injustice* – commemorating the twentieth anniversary of Harvey's book – Merrifield and Swyngedouw (1996) urge critical thinkers to "...rethink the relationship between spatiality, power and justice, but also (...) of socially just urban policies" (p. 3).

Ed Soja (2010) became the principal advocate of spatial justice in his book *Seeking Spatial Justice* where he defines it as "...a spatial expression that is more than just a background reflection or a set of physical attributes to be descriptively mapped". To make his point clear he continues:

> Calling it spatial justice is not meant to imply that justice is determined only by its spatiality, but neither should spatial justice be seen as just one of many different components or aspects of social justice
>
> (p. 5)

Soja is well known as a theorist arguing that space is social anyway – that there is no need to differentiate between social and spatial processes, and he insists on "putting space first". I think he went too far in using spatial justice as an umbrella term, following the so-called "spatial turn" to unify all aspects of justice. Therefore, I prefer the term socio-spatial justice/injustice to avoid misunderstandings coming out from spatial verification and to include multiple and multi-scalar forms of justice/injustice as inherently social and spatial. In what follows, I discuss the socio-spatial injustices derived from inequalities at the Eurozone scale and at the national/regional scale in SE countries.

The unfair terrain of the Eurozone and the hypocrisy of financial solidarity

The bailout for Greece in March 2010, and those for Ireland in December 2010, Portugal in 2011 and Cyprus in 2013, can hardly be taken as major solidarity steps within the EU, or even globally if the IMF is included. The austerity packages imposed on southern countries and Ireland were not a "just policy for Southern people", as argued by many European political leaders, the press and the then Greek Prime Minister George Papandreou.

Ostensibly, the bailouts were supposed to rescue these countries, yet there was not a single cent in these packages that was aimed at helping Southern people or any city and region. Instead, the bailouts were and are to provide a guarantee against a debt default by the four, restoring the balance sheets of Europe's banks holding Southern bonds. They were therefore entirely a bailout for holders of Southern government bonds, i.e. the major European banks, and the rescue plan was to their benefit. Southern states, regions and the people were written off on the basis that their banks were already net debtors. Worse still, Southern people were forced to accept fiscal contraction, in the case of Greece, involving huge cuts to public spending, services and welfare payments and job losses rising to circa 1 million by the beginning of 2011, or 23 per cent of the workforce. In Spain, the government approved an austerity budget for 2011 that included a horizontal tax rise and 8 per cent spending cuts. In Italy, similar austerity packages of 70 billion euros were introduced by 2011 on top of an austerity package introduced the previous year. Unlike Greece, neither of these countries was under Troika supervision.

On a broader front, SE countries are becoming a test case for the latest phase of neoliberal financial correction, making them "states of exception" in the wake of the economic and financial crisis. These policies are among the most severe and unjust fiscal austerity packages in Europe since the Second World War. In fact, they are not only unjust policies but are a radical attack on politics and are the preeminent expression of the neoliberal hatred of democracy.

The economist Pablo Beramendi (2012), in his book *The Political Geography of Inequality, Regions and Redistribution,* argues that variegated fiscal structures in different political models of administration are the main factors causing inequality. Fiscal structures determine interpersonal and interregional redistribution. Although he limits himself to personal income and GDP as inequality indicators, he makes the important point that changes in economic geography and political representation play key roles in eliminating inequality. Writing about the EU, he describes it as having a decentralised fiscal structure and a heterogeneous economic geography that results in:

> … a very narrow window for redistributive efforts and inequality reproduces itself by contributing to the selection of institutional arrangements that, in turn, will protect existing territorial inequalities
>
> (p. 13)

A major problem in the Eurozone is the spatial representation of citizenry. It does not refer to "Europeans" but only to Germans, French, Greeks, Spaniards, Irish etc., which in the first crisis reproduced nationalistic and antagonistic attitudes, far from the ideas of "United Europe" and "European solidarity". Old prejudices and geographical imaginations, discussed in the previous chapter, were mobilised to rationalise the Southern subaltern position and to strengthen the image of the "proper European" as a Northerner. This socio-spatial representation works for the benefit of European elites because it homogenises social classes and places and hides the connection between internal and external class forces, which is mediated by the national scale.

A major difference with previous global crises and the present one is the framing of the scale on which it is produced and experienced, as was shown in Chapter 3. In the new form of multi-scalar governmentality of the EU, and more closely in the Eurozone, the destabilisation and rescaling of the national territorial scale is fundamental because at this scale, social protection takes place. The wider rescaling of the EU through the new governance regime, and the restructuring of capital circulation and accumulation, leads to new forms of political exclusion and socio-spatial injustice. Eric Swyngedouw (2000: 70), writing on globalisation, emphasises how rescaling undermines democracy and citizenship rights:

> ...The double rearticulation of political scales (downward to the regional or local level; upward to the EU, NAFTA, GATT, etc; and outwards to private capital) leads to political exclusion, a narrowing of democratic control, and, consequently, a redefinition (or rather a limitation) of citizenship rights and power

The creation of a monetary union without a broader European social protection equivalent, and without any prospect of creating one, sheds light on the question of why neoliberalism is still unchallenged after so many years of austerity. Following Nancy Fraser (2008), while those who established the euro were trying to overcome the "Westphalian" political-geographical imagination of bounded national spaces of the former nation-states, the everyday operation of the Eurozone, and particularly the management of the crisis, seems to re-establish it. This has added more confusion to the justice discourse, which continues to operate under the "Westphalian" framework. Who and where count today as subjects of justice in the Eurozone?[5] In the old "Westphalian" national framework, the "who" and the "where" of justice were matters encompassing economic redistribution, legal and cultural recognition and equal political representation, all of which are underpinned by the normative principle of participatory parity (Fraser, 2008). None of those who participate today in the Eurozone enjoys these conditions *at the Eurozone level*. They have moral standing as subjects of justice only in their sovereign states, and this is a major contradiction that surfaced during the present crisis.

The introduction of the euro, along with other instruments, is used as a mechanism through which global capitalist pressures are shifted onto regional labour markets in order to secure capital's profitability, as analysed in Chapter 3. European labour markets, however, due to weak geographical mobility of labour, remain a national/regional affair and are still regulated by national laws. In SE, the Troika got things exactly the wrong way around from an economic policy viewpoint, but perfectly well from its neoliberal class strategy. It has focused on labour cost, when labour costs are among the lowest in Europe, and this policy produces major socio-spatial injustices. Squeezing workers over six years did not increase competitiveness and has merely had the effect of further limiting their life expectancy – in Greece by 1.5 years – and eroding the countries' tax bases – in Portugal by 9 per cent. The euro is now a source of discord, as both nominal winners and losers feel that they are at the mercy of Brussels and Berlin: the Southerners complain that they have been occupied, the Germans that they are being blackmailed. So, despite efforts by union federations to organise themselves at the EU level, national/regional spaces are the prime fields of struggle between capital and labour. This is another major spatial/scalar contradiction in the process of European integration, which, again, financial capital exploits for its benefit.[6] While in principle the Eurozone is the space in which the euro's class function will be judged, in reality this takes place at the national and regional scales, hence the necessity to smash the resistance of Portuguese, Spanish, Italian and Greek labour as a test for the rest of Europe and to blame the Southerners as lazy.

Under the ideological hegemony of neoliberalism, wealthy and middle-class citizens in the richer regions of Germany, the Netherlands, Austria, Finland and other countries have no incentive to agree to a more just fiscal structure that transfers resources to other areas of the union. Poor citizens in rich regions prefer redistribution within their own regions rather than engaging in class solidarity with the poor people in other regions of the EU. On the contrary, poor citizens in poor regions in Southern and Eastern Europe, who are the recipients of structural funds, support a stronger central mechanism of redistribution that does not exist yet in the EU and the Eurozone.

The discussion so far highlights why an application of socio-spatial justice at the Eurozone level seems extremely difficult to attain. The spatial/scalar displacement of citizenry, justice and class struggle makes it difficult, for the time being, for some citizens of the Eurozone to accept actions of solidarity with those who "failed" in the South. It makes it also difficult for some unions and ordinary people to understand the problems of "others" as "their own", and in turn to organise and synchronise their struggle at the EU level. The opposite is true for the various factions of the left, organised in political parties or in independent non-parliamentary groups and for radical unions. During all these years of austerity, they provided invaluable support and solidarity, practical as well as moral. In the following Chapter 6, their actions are discussed in detail.

Socio-spatial injustices at the national and regional scale

During the decade before the crisis, there was no shortage of big promises by national and EU elites. While a full attack by neoliberal policies across Europe was and remains in place – always framed as modernisation and efficiency – we heard promises such as "a new European century" (E. Barroso, ex-President of the EU and now high executive in Goldman Sachs), "we don't let business dominate our advisory groups" (EC press release), "we will deliver smart growth, jobs and social security" (A. Samaras, Greek ex-Prime Minister) and many more. At the same time, hundreds of thousands of jobs were destroyed, tax cuts for the rich didn't "trickle down" to create jobs for the masses, the manufacturing sector was devastated (with the exception of Germany and its satellites) and the deregulation of the financial sector followed extended privatisations everywhere. The immediate effect was an increase in social and spatial inequalities and the uneven hollowing out of the working and middle class across countries and regions.

When the Troika's technocrats arrived in SE to apply austerity measures in collaboration with national elites and domestic capital, with cuts to employment, wages and social welfare, they found an already weak welfare system and exhausted citizens. In the past, extended family support networks and kinship had compensated for the traditionally weak welfare systems in SE. This repository of informal protection dissolved due to horizontal cuts in income to all families and the elimination of multiple employment opportunities. The model of at least one fully employed person per family while other family members had temporary or insecure jobs changed to dependence mainly on the pensions of the family's elderly members (Andreotti and Mingione, 2014).

A study financed by EU (DG) Directorate-General Research, with the title "Why socio-economic inequalities increase? Facts and policy responses in Europe" (see, Perrons, 2010), argues that economic policies in the EU are inadequate to handle increasing inequalities and in many cases have fuelled them. It uses the Gini coefficient as an illustration of income inequality and, focusing on SE, depicts a considerable increase in inequalities in Portugal and Italy until the mid-2000s, with a relative decrease in income inequality (see Table 5.1) in Greece and Spain. These tendencies reversed in 2014 with dramatic increases in Greece and Spain, while in Portugal and Italy, income inequality decreased.

In 2014 in the EU, only the UK had a higher Gini coefficient (0.348) while the lowest was in Denmark (0.249) and the EU average was 0.299. In

Table 5.1 Income inequality in Southern Europe 1980–2014, Gini coefficient

	Mid-1980s	*Mid-1990s*	*Mid-2000s*	*2014*
Greece	0.336	0.336	0.321	0.340
Italy	0.309	0.348	0.352	0.327
Portugal	0.329	0.359	0.385	0.338
Spain	0.371	0.343	0.319	0.335

Source: OECD (2008, 2015), where 0=complete equality, 1=complete inequality.

the same year, the distribution of total wealth towards the 10 per cent of wealthier individuals/households was 40 per cent for Greece, 45 per cent for Spain, 48 per cent for Italy and 53 per cent for Portugal – another indication of income inequality concerns earnings compared to productivity. In OECD countries, during 1996–1999, 48 per cent of the workforce saw their earnings rise more slowly than productivity. Between 2003 and 2006, the gap widened to include 61 per cent of the workforce (OECD, 2006). Employment does not provide a guarantee against poverty anymore. The research looked also at income inequalities spatially. Inequality between member states generally decreased but inequality between regions and within regions increased.

A 2016 report by Credit Suisse (Global Wealth Data book, 2016) provides additional information on the uneven and unjust distribution of national wealth among households during the years of austerity. It shows how the richest 10 per cent of the population in the four countries of SE got richer during the crisis years and owned more than 50 per cent of national wealth in each country. Austerity and the Troika's interventions are very class based, indeed (see Box 5.2).

Box 5.2 How the rich get richer during the crisis

The bourgeois mantra "crises create opportunities" finds application in the distribution of national wealth among rich households during the crisis. The Credit Suisse report establishes wealth inequality measured by the share of the wealthiest 10 per cent of adults as compared to the rest of the adult population. Net wealth is defined as the value of financial assets plus real estate assets owned by households minus debts. Private pension fund assets are included, but not entitlements to state pensions.

According to the report, the richest 10 per cent of households held in 2010 38.8 per cent of national wealth in Greece, 45.7 per cent in Italy, 52.7 per cent in Portugal and 45.5 per cent in Spain. Greece, at the beginning of the crisis, looked less polarised than the other countries. By 2016, the 10 per cent got richer in all countries and in Greece much richer compared with a few years earlier. Thus in 2016, the 10 per cent in Greece held 54.0 per cent, in Italy 54.7 per cent, in Portugal 58.8 per cent and in Spain 56.2 per cent. Among the rich, there are further polarizations and inequalities: In 2016, the 1 per cent in Greece held 24.3 per cent of the total wealth, in Italy 25.0 per cent, in Portugal 28.0 per cent and in Spain 27.4 per cent. Although these figures are lower than the 1 per cent in the USA in 2014 that owned 35 per cent, nevertheless, they show the class character of the crisis and how some people manage to get comparatively richer.

In terms of personal wealth, the number of millionaires – people owning more than 1 million USD – increased during the crisis, while many millionaires left their countries to head to the UK and the USA. By 2016, Spain had

193,000 millionaires – of which 386 were billionaires, Italy 295,000 – of which 1,132 were billionaires, Portugal 54,233 and Greece 55,000, both with a dozen billionaires. A few years earlier, Spain in 2008 had 96,500 millionaires, Italy in 2004 had 188,000, Portugal in 2012 had 10,750 and Greece in 2005 had 38,000. All the above, in relation to the devastating effects on the 90 per cent of the population in the four countries, provide good ground for indignation and revolt, as described in Chapter 6.

Source: credit-suisse.com/articles/2016/the global-wealth-report; www.onthepulse.es/spanish economy-news; www.theportugalnews.com; www.wantedincome.com; Forbes (2016).

Regional unevenness in the EU since the 2000s is also increasing and highly correlates with particular EU policies. During the early euro years, 2000–2013, all development policies of the EU, including regional policies, were subsumed under the Lisbon strategy. It was assumed that regional competitiveness, as the main development axis, would provide "growth and jobs". All EU regions became eligible for funding, and although extra funding was available for the less developed regions, those in North-Central Europe grabbed the lion's share, and in the end, this policy shift was catastrophic for southern regions (Thoidou and Foutakis, 2006). The sequel to this story is Europe 2020, a strategy promoting "smart, sustainable, inclusive growth". By then, everything in Europe is supposed to be "smart" to "improve the business environment". In addition, all policies should follow neoliberal "fiscal discipline" and finally, "Fiscal consolidation and long-term financial sustainability will need to go hand in hand with important structural reforms, in particular of pensions, health care, social protection and education systems" (Europe, 2020: 24).

A major problem contributing to social and spatial inequality is the rise of atypical and informal work in all EU regions, but particularly in the South. Shorter hours, temporary work, work-on-call etc., in short precarity, has expanded, and these jobs produce lower incomes and contribute less to taxes and to insurance/pension funds, thus contributing further to public debt. Differences between workers with atypical/informal contracts and permanent contracts from 2000–2009 (i.e. before the crisis), increased in Greece by 26.4 per cent, in Portugal 21 per cent, in Italy 18.8 per cent and in Spain 16.7 per cent, while in the whole of the EU it increased by only 1.9 per cent (Eurostat, 2010). This rise of non-standard forms of work contributed to an increase in rates of employment for women, while gender inequality increased due to lower pay, poorer conditions of work, poorer pension schemes and the lower availability of social services for women with children (Perrons, 2012a; Karamessini, 2013; Vaiou, 2014). Since women account for high percentages of jobs in the public sector, any change in terms of employment or reductions in wages affects women more than men (Karamessini and Rubery, 2013). Finally, the housing crisis in urban areas – mainly in Spain and Italy – particularly affects single mothers, as the cost of childcare increases due to the reduction in public spending. Austerity widens gender inequalities.

Gialis and Leontidou (2016) measured the regional effects of crisis in the amount of flexible and informal employment in Greece, Spain and Italy. They found a considerable increase in Greek regions, followed by Spanish and Italian regions. Their findings contradict the Troika's beliefs around persisting rigidities in SE labour markets. Subsequent policies to reduce wages and to eliminate the supposed job protection in the South did not increase competitiveness but instead worsened the situation.

On the regional development theoretical front, the hegemony of neoliberalism has succeeded so far in defending the taken-for-granted mainstream model of the competitive and entrepreneurial city/region, thus undermining any attempt at welfare planning. The entrepreneurial city/region, competition among cities and regions, city marketing, regional branding, etc. has taken over the reins of planning and policymaking. One immediate effect of this replacement is the intensification of regional inequalities. By the same token, any discussion of inequalities, socio-spatial justice and solidarity has also been practically erased from the European regional development agenda.[7] The Sixth Report on Economic, Social and Territorial Cohesion (2014) discusses regional disparities in EU regions (see Box 5.3)

Box 5.3 Crisis and austerity policies increase socio-spatial inequalities

Extracts from: *EU, Sixth report on economic, social and territorial cohesion*, Brussels 2014:

> ...Since 2008, public debt has increased dramatically, income has declined for the majority of people across the EU, employment rates have fallen in most countries, unemployment is higher than for over 20 years, while poverty and social exclusion have tended to become more widespread. At the same time, regional disparities in employment and unemployment rates have widened as have those in GDP per head in many countries, while in others they have stopped narrowing.
>
> ...In 2000, average GDP per head in the most developed 20% of regions was about 3.5 times higher than that in the least developed 20%. By 2008, the difference had narrowed to 2.8 times.... However, the crisis seems to have brought this tendency to an end and between 2008 and 2011, regional disparities widened." "...Some regions have been hit severely, others hardly at all. This is particularly evident with regard to regional unemployment rates. In 2008, five regions had an unemployment rate

above 20%. In 2013 the number had increased to 27%. At the same time, unemployment has gone down in many German regions because of the relatively strong performance of the German economy since the global recession in 2008–2009.

...On average, public investment within the EU declined by 20% in real terms between 2008 and 2013, in Greece, Spain, Portugal and Ireland, by over 60% and in the EU-12 countries, where Cohesion Policy funding is particularly important, by 32%" "...Higher risk of poverty or social exclusion is another legacy of the economic crisis. There are now around 9 million people at risk of poverty or exclusion in the EU, the increase being particularly pronounced in Greece, Spain, Italy and the UK. A key issue is the variation within countries".

www.ec.europa.eu/regional_policy/sources/docoffic/official/reports

Descriptions in the EU DG Research (2010), in the EU Sixth Report on Economic, Social and Territorial Cohesion (2014), in the book by Karamessini and Rubery (2013) and in many journalistic reports do not constitute an unfortunate and shocking exception, but rather a critical regularity imposed by financialisation and austerity policies (Athanasiou, 2012). The Troika's intervention is not only an unconstitutional act – therefore illegal and unjust – but works also as a "collective punishment" imposed on people for actions and policies for which they were not responsible for (in other words, scapegoating). In this respect, I want to argue that in SE there is a direct correlation between increases in socio-spatial inequalities and injustice. The slogan in demonstrations across SE "the crisis should be paid for by bankers and politicians" makes the point clear and explains the widespread anger and the feeling of injustice among those who mobilised.

The German Bertelsmann Stiftung, an influential conservative institution, in two reports in 2014 and 2016, made direct reference to social justice with a cross-national comparison in the EU, documenting socio-spatial justice at the national scale (see Schraad-Tischler and Kroll, 2014 and 2016). The 2014 report argues that:

> ...Social injustice clearly increased in recent years, most obviously in the crisis-battered southern European countries of Greece, Spain and Italy, as well as in Ireland and Hungary (and to a lesser extent in Portugal). Social security systems have been badly undermined by austerity measures.
>
> (p. 6)

Greece, according to Schraad-Tischler and Schiller (2014, 2016), has the lowest social justice index score[8] among the crisis-ridden countries in both reports: in 2014 3.57, (2013 data) and in 2016 3.66 (2015 data). With 50 per cent youth unemployment in 2016, a health and educational system badly undermined by austerity, a significant increase in poverty among children and pensioners and with the highest public debt, the future looks bleak for coming generations, a situation that is of no surprise. At the bottom third of the social justice index, after Greece, are also (in rank from 3.66 to 5.03) Romania, Bulgaria, Spain, Italy, Hungary, Portugal, Latvia and Cyprus. The highest score is for Sweden with 7.51 (2016 Report).

Both reports, however, restrict themselves to data description, avoiding any critical comments on the harsh austerity. An innocent reader of these reports may conclude that national governments with their free will have applied austerity. Consequently, only endogenous factors should be blamed and that the "...social security systems have been badly undermined" by national governments alone. Although national political forces in power during the crisis passed austerity measures in parliaments, continuous intervention and surveillance by the EU, the ECB, the Eurozone and the Troika should also be taken into account. Thus, I read socio-spatial injustice in these reports as the combined effect of multi-scalar processes of spatiality and power imposed by the Troika and national governments.

Two important dimensions of injustice, according to the reports, are the ratio of people to the total population at risk of poverty and the ratio of people to the total population in severe material deprivation (see Tables 5.2 and 5.3). While in all SE the risk of poverty increased after 2008, with a record high in Greece in 2015, the EU average declined after 2014. Severe material deprivation[9] increased in all SE countries until 2014, but Portugal and Spain had lower ratios compared to other SE countries and below the EU average in 2014. Only in Greece did deprivation increase, to 22.2 per cent, in 2016.

Poverty and material deprivation, however, are not solely an SE characteristic. Bulgaria in 2014 had the highest score with 49.3 per cent of population at risk of poverty[10] and the highest score with 45.9 per cent of total population facing severe material deprivation. By 2016, the lowest was in the

Table 5.2 Risk of poverty, % of total population

Rank of country (2014)	*2008*	*2011*	*2014*	*2015*
16 Portugal	25.0	24.9	25.3	26.6
EU average	24.5	23.7	25.4	23.7
18 Cyprus	25.2	23.5	27.1	28.9
19 Spain	23.3	24.5	27.3	28.6
20 Italy	26.0	24.7	28.4	28.5
21 Ireland	23.1	25.7	30.0	–
25 Greece	28.3	27.6	34.6	35.7

Source: Bertelsmann Stiftung (2014); Eurostat (2016).

Table 5.3 Severe material deprivation, % total population

Rank country (2014)	*2008*	*2011*	*2014*	*2016*
9 Spain	3.5	4.5	6.2	6.4
18 Portugal	9.6	9.1	10.9	9.6
EU average	11.0	9.7	11.5	–
20 Italy	6.8	7.0	12.4	11.5
23 Cyprus	13.3	9.5	16.1	15.4
24 Greece	11.5	11.0	19.5	22.2

Source: Bertelsmann Stiftung (2014 and 2016).

Netherlands, France and Denmark. In Spain, according to the report, the crisis affected poverty in different age groups quite differently. While "only" 14.5 per cent of older people are at risk of poverty and only 2.7 per cent suffer severe material deprivation, the corresponding percentage for children is twice as high.

General unemployment was between 5 per cent and 10 per cent from the mid-1980s until the beginning of the crisis in 2009 for all SE countries except Spain (see Figure 5.1), close to the EU average and Germany. The housing bubble contributed to a sharp decrease in unemployment in Spain from the late 1990s until 2007, but after 2009, together with Greece, it climbed above 25 per cent, the highest in the EU in a span of 30 years (the increase for Greece was a massive 20 per cent in just four years). Italy managed a modest increase during the 2009–2014 crisis, while by the end of it, Germany had achieved the lowest unemployment figures since 1991 at 4.9 per cent. The difference between Spain and Greece and the median figure for the EU, not to speak of Germany after the imposition of austerity, is striking. Although it is an aggregate figure that cannot accurately depict localised unemployment, which is highly spatially uneven within countries, it nevertheless highlights in the crudest way the socio-spatial injustice between the countries in question.

Regional employment and unemployment rates in general follow the corresponding national pattern. Thus, the highest regional employment rates were recorded in German, Finish, Swedish and UK regions, while the highest unemployment rates were in Greek, Italian and Spanish regions (Eurostat, 2016). The highest employment records tended to be recorded in southern regions of Germany and Sweden, above 75 per cent of population, and the lowest in Southern Italy (Calabria, 42.1 per cent and Campania 43.1 per cent) and Dytiki Ellada (47 per cent) in Greece. In general, among the 34 EU regions where the employment rate was below 60 per cent, the vast majority of unemployed were concentrated in Greece, Spain and Italy. Geographical unevenness also characterises the internal structures of Spain and Italy with high employment rates in the North and high unemployment rates in the South, that is in Andalusia and the Mezzogiorno respectively. These disparities are further intensified at the local scale by considerable gender gaps,

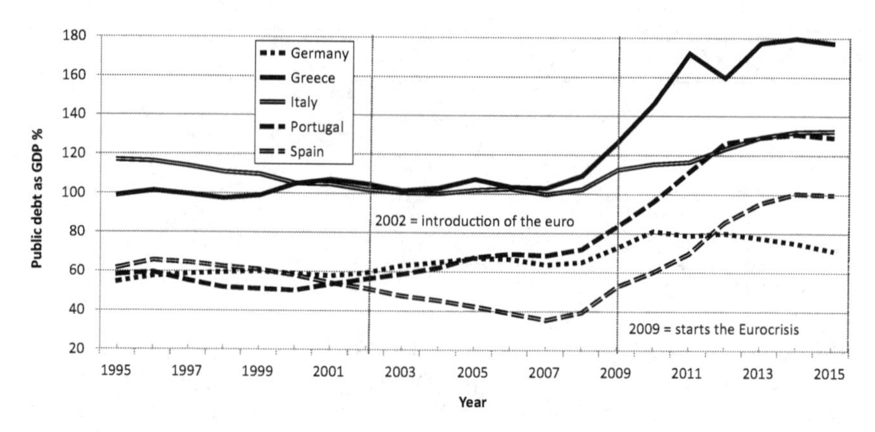

Figure 5.1 Unemployment in the European Union, Germany, Greece, Italy, Portugal and Spain, 1985–2015.

Source: "Unemployment rate" in *Annual macro-economic database (Ameco)*, (2016).

with male employment rates being 20 per cent higher than women in Voreio Aigaio in Greece, in Campania and Basilicata in Italy and in Extremadura and Murcia in Spain. Finally, the gap between urban and rural areas in terms of employment is also evident, with urban areas in the South having the highest rates in cities. In terms of long-term unemployment, in four Greek regions (Peloponnisos, Attiki, Sterea Ellada and Dytiki Ellada), in 2015, more than three quarters of the unemployed had been without work for at least a year. The same is true for seven Italian regions, three regions in Portugal and four regions in Spain. Finally, current policies in the EU regarding innovation in industrial sectors, where product innovation and capital intensity is high and labour intensity is low, may reinforce socio-spatial inequality in particular SE regions due to the limited contribution to general employment.

Youth in all four SE countries have become the stereotype of the "lost generation", living precariously, unable to get a job, with earnings barely enough to pay the rent, and of course without hope of retirement and a pension. The tendency to return to parents' or to grandparents' homes increases geometrically parallel to homelessness. In terms of youth unemployment (see Table 5.4), it is of little surprise that Spain and Greece fall at the tail end of the ranking, and this explains in part their anger and frequent militant revolt, to be discussed in the next chapter. Another important dimension of injustice is the increased number of early school leavers, as young people don't see any employment future and must try to make a precarious living in the informal sector (see Table 5.5). This is the only indicator in which Greece has better figures than the other three SE countries. In Greece and Spain, at the regional level, the highest numbers of school leavers are in the tourist areas and large urban centres.

Table 5.4 Youth unemployment as % of total population below 30 years

Rank country (2014)	*2008*	*2011*	*2014*	*2016*
EU average	15.2	23.2	26.2	–
18 Ireland	12.7	27.6	26.8	20.9
23 Portugal	16.4	22.4	37.7	32.0
25 Italy	21.3	27.8	40.0	40.3
27 Spain	24.5	41.5	55.5	48.3
28 Greece	22.1	32.9	58.3	49.8

Source: Bertelsmann Stiftung (2014 and 2016).

Table 5.5 Early school leavers[11]

Rank country (2014)	*2008*	*2011*	*2014*	*2016*
11 Ireland	11.3	11.5	8.41	6.9
12 Cyprus	13.7	12.7	9.1	5.3
19 Greece	14.8	13.7	10.1	7.9
EU average	13.4	12.1	10.4	–
24 Italy	19.7	18.8	17.0	14.7
26 Portugal	35.4	28.7	19.2	13.7
28 Spain	31.7	28.2	23.6	20.0

Source: Bertelsmann Stiftung (2014 and 2016).

Regional youth unemployment in 2015 was concentrated in those regions that experienced relatively high overall levels of unemployment. In all regions of Greece and Spain, and in Southern Italy, youth unemployment was above 40 per cent. Furthermore, in eight regions in Greece, seven in Spain and five in Italy youth unemployment was higher than 50 per cent. An OECD (2016) study calculated regional variation in youth unemployment that depicted high gaps in Italy (16 per cent lowest, 60 per cent max), in Greece (34 per cent to 70 per cent), in Spain (45 per cent to 65 per cent) and in Portugal (28 per cent to 50 per cent).

Also of little surprise is that the lowest youth unemployment rates for 2016 are in Germany and Austria, two destination countries (together with the UK, the Netherlands and Sweden) of the "brain drain" coming from SE. Germany is the leader in the prevention of youth unemployment, followed by Austria, the Netherlands, Malta and Denmark, all below 13 per cent. In the 1960s and 1970s, the countries of SE were traditional regions of emigration, sending thousands of people as workers to Western Europe, as discussed in Chapter 2. Today, high unemployment among young people, particularly university graduates, results in "brain-drain", what Lois Labrianidis (2011) calls "investing in leaving", with high costs for SE countries and consisting of an issue of intergenerational justice (see Box 5.4).

Box 5.4 Brain drain from SE

Labrianidis (2011) estimated that during 2006–2011, i.e. even before the crisis, 160–180,000 graduates left Greece; half of them with a PhD. Similar brain-drain takes place in Spain and Portugal, undermining potential recovery capabilities in the future (see also Labrianidis and Vogiatzis, 2013). Another research by McKinsey Global Institute (*Kathimerini*, 4 December 2016, in Greek), estimates that between 2008 and the first half of 2016, i.e. including the crisis period, 350,000–430,000 people left Greece, two thirds of them with a university degree. In economic terms, the same institute estimates that this brain drain contributed 12.9 billion euros in GDP and 9.1 billion in taxes to the destination countries, mainly to Germany, the UK and France. More than 9,000 Spaniards, 4,700 Italians and 2,000 Greeks took intensive German language classes offered by the Goethe Institute in 2015 alone.

The lost income for Greece is even higher if we add the costs of social reproduction of these people and the public expenditure for their education, estimated at 8 billion euros. An indication of the future difficulties facing SE countries is the current (in 2016) inability of employers to find qualified personnel for their vacant posts: 33 per cent in Greece, 22 per cent in Spain, 20 per cent in Italy and 18 per cent in Portugal.

Perhaps the clearest act of intentional socio-spatial injustice was the attack on National Health Systems (NHS). Austerity programmes represent an immediate threat to the survival of NHS everywhere – as in the UK under Thatcher and Blair and nowdays in the USA – but in SE have reached extreme levels. They constitute a direct bio-political intervention, a practice of "necessary exclusion", according to Troika's technocrats, of poor social groups, places and regions from the most vital social service enjoyed by other European citizens. The average annual reduction in per capita health expenditure during the period 2009–2012 was 11.8 per cent in Greece, 7.9 per cent in Portugal, 4.3 per cent in Spain and 2.4 per cent in Italy (Petmesidou, et al., 2015). Since 2012, 7 per cent of Spanish pensioners, i.e. 662,000 people, cannot afford to buy medicines because the neoliberal government increased their contribution to buying them. These reductions particularly hit the treatment of cancer, and, in combination with diminishing access to healthcare for economic and accessibility reasons, this means that the number of healthy life years has significantly diminished in all four countries, while in the EU as a whole it remains stable for both women and men (see Table 5.6).

After six years of austerity, Greece's public healthcare system has been enormously damaged, such that "Doctors Without Borders" describe the situation as a "humanitarian crisis". The UN circulated in 2011 an

Table 5.6 Mean number of healthy years, 2009–2013, total change

	Greece	*Italy*	*Portugal*	*Spain*	*EU*
Women	−2.2	−3.6	−2.8	−0.6	0.0
Men	−0.9	−3.6	−0.6	−0.7	+0.1

Source: Eurostat (2014).

independent UN expert's warning that austerity measures could result in violations of Southern peoples' human rights, particularly for Greeks, such as "...rights to food, water, adequate housing and work under fair and equitable conditions" (UN, 30 June 2011). The Greek neoliberal government used the Troika's proposal for cuts as an excuse to introduce a regulation in 2012 that those who became unemployed, i.e. 1.5 million people by 2013, together with the loss of their job, lost access to the Greek NHS. The immediate effect was that 13.6 per cent of Greeks could not afford to receive medical care or treatment, dental care (18.2 per cent) or mental health care services (4.2 per cent), while 12.5 per cent did not have the financial means to buy medicines prescribed by a doctor (ELSTAT, 2013, health survey). This inhuman, undemocratic and unjust measure was revoked in 2016 by the coalition government lead by SYRIZA, although it could not change the predicted life expectancy at birth, which has decreased by three years. Similar figures for the reduction in life expectancy exist in Portugal, 1.5 years and in Spain, 1.2 years, while long forgotten diseases such as tuberculosis are on the rise again (*Efimerida ton Syntakton*, 4 May 2013, in Greek). In addition, during 2011–2012, suicides in Greece, mainly for economic reasons and despair, rose by 45 per cent and in Italy by 36 per cent (*Der Spiegel*, 15 April 2012). Finally, in Spain and Greece, we see another major dimension of injustice: the "energy poverty" (Vatavali and Chatzikonstantinou, 2015). Due to high unemployment and low benefits, many families do not pay their utility bills and companies cut electricity, gas and water supplies. In Barcelona, representatives of Caritas argue that thousands of families live under these conditions and urge the municipality to take immediate action. Newspaper articles report incidents of people, particularly elderly women, who die from fire accidents (six in Greece, three in Spain in one year) trying to keep warm with old-fashioned wooden stoves and using candles at night. (*Efimerida ton Syntakton*, 25 November 2016, in Greek). Austerity kills! Literally!

As national systems of social care break up, the burden of care shifts to families, increasing the pressure on women and on several bottom-up solidarity initiatives, discussed in the next chapter. The cost of social reproduction is externalised, and, at the same time, it is domesticated, adding another dimension to socio-spatial inequalities among households (Vaiou, 2014).

We know that capital tends to periodically destabilise the conditions of social reproduction, but what happens in SE seems extreme and also contradictory. Capital accumulation needs a healthy, well-educated and trained labour force, and this is precisely what is at present being destroyed (Fraser, 2016). The latter explains in part the reaction of employers' organisations in Greece, Spain and Italy against some of the cuts in wages and pensions in the three countries (*Avgi* newspaper, 12 November 2015, in Greek).

For the authors of the Bertelsmann Stiftung reports, the concept of justice "...is concerned with guaranteeing each individual genuinely equal opportunities for self-realization through targeted investment in the development of individual 'capabilities'" (p. 13) – a very conservative definition indeed. The direct reference later in the report to Amartya Sen (2009) and their demand for strong state intervention cannot alter the limitations of their conceptualisation of justice. The most important element of injustice in the first place is the provision of equal opportunities to structurally unequal individuals in terms of class, gender, location/spatial reference, ethnicity and sexual preference. Second, the emphasis on individuals and not on social groups, social classes and particular places is indicative of the non-critical attitude of the report. Equality among individuals under neoliberalism is anyway questionable, but even if it could be materialised, it spawns inequalities among social groups and places. As a result, while the report illustrates the negative social and economic outcomes of austerity measures, it refuses to provide even a modest critique of austerity policies. As the section on Greece argues, accepting austerity is necessary because it reduces the cost to business:

> ...the terms of the bailout (in Greece) increased unemployment and disabled government policies for helping people into work. At best, this is a flexibilization of the labour market that will reduce costs and increase competitiveness, allowing a more sustainable economic path in the future. But in the short and medium term, such austerity simply increases unemployment dramatically
>
> (Sotiropoulos/Featherstone/Karadag, 2014, available at www.sgi-network.org)

There is, however, a different reading of these data and tables. Three decades of neoliberal policies in SE and seven years of harsh austerity measures are enough to destroy, with variations among countries and regions, what is left of the welfare state and welfare planning. Without redistribution policies, socio-spatial inequalities increase and so do indices of injustice. All these years, national governments in SE and the EU keep declaring that their aim is to reduce social and spatial exclusion – the famous social and territorial cohesion strategy – while their practical politics create more of the same. For example, during the crucial period 1999–2007, structural funds in SE were only conditionally effective. They had a significant positive impact on total employment only in cases of a low share of low-skilled population. In this respect, and despite the high per capita allocation of funds at national scale, the majority of SE regions having low-skilled specialisations did not benefit from these policies. Promoting regional and urban competitiveness, as have all EU policies since 2000 and theories of the "Third Way", inevitably ends up with uneven distribution of wealth, education and health conditions among people, regions and countries. Those who accept the above believe that the greed of capital is good because it promotes investments leading to growth. Elites at the top often look down, sometimes with fear, at those excluded and use prejudices to defend/secure their elite position. And all seem to ignore that true socio-spatial justice, as Dorling (2010: 11) insists "...will both create and require much greater equality than is as yet widely accepted".

* * *

Summing up, progressive "Third Way" economic geography and regional development slide towards mainstream de-politicising paths, and researchers and policy makers have succumbed, perhaps unintentionally, to the charms of neoliberalism. They had idealised flexible specialisation in small firms and the success in SE industrial districts and did not recognise warnings about crises in their model regions. It was ironic that when EU regional policies integrated these ideas, their model regions were already in trouble. The crisis in SE and in the Eurozone found "Third Way" and neoliberal models operating on the same track so that both promoted policies for a competitive and entrepreneurial city and region. Furthermore, during the programme period 2007–2013, i.e. during the crisis, the distribution of structural funds to EU regions followed criteria based on pre-crisis data, and many SE regions were excluded. Thus, at the time that southern regions needed more than ever the support from the structural funds, the sudden fall in GDP and the dramatic increase in unemployment were not taken into account. The failure to address the wider "state of exception" imposed in SE, directly in the form of memoranda, as in Greece and Portugal, and indirectly as in Spain and Italy, is perhaps the largest failure of regional policies after 2013 reiterated by the Bertelsmann Stiftung reports, and this is true for the majority of approaches to the SE crisis. In addition, the wealth inequalities depicted in the Credit Suisse report showed how the rich get richer during the crisis and highlighted the class character of austerity policies.

The state of exception establishes a hidden but fundamental relation between conditions of socio-spatial justice and their absence: Laws and justice are in place but devoid of any obligation to be implemented. Under these conditions SE citizens do not count as subjects of justice to the same degree as the other EU citizens. During the last seven years in SE we haven't enjoyed economic redistribution, cultural recognition, legal and social protection at different spatial scales, or as Fraser (2016) puts it, distributive justice, justice of recognition and representational justice. The precarity regime becomes a performative force for legitimising state violence, unemployment, poverty and the reduction of health standards. In short, SE citizens are turned into mute bearers of what Giorgio Agamben (1998) calls "bare life", deprived of their right to a better future.

Notes

1 I refer here to influential senior researchers responsible for drawing up theories. There exists dozens of others who made important contributions related to crisis, geography and planning in Europe and are referred to in other parts of this book. In addition, different groups of younger researchers are engaged in promising research and activism across Europe, particularly in the South. Among the many workshops and conferences they have organised are: "Crisis regimes and emerging social movements in Southern Europe: urban development, housing and local struggles", Encounter Athens, Athens February 2013; "Crisis-Scapes-Athens and beyond", Athens May 2014; "Urban Austerity: impacts of the global financial crisis on cities in Europe", Weimar, December 2014; "Local Resistance, Global Crisis: Solidarity and Left Politics for the 21th Century", Maynooth, June 2014; "Urban Conflicts Seminar", Thessaloniki, June 2015; and "From Contested Cities to Global Urban Justice: Critical Dialogues", Madrid, July 2016.

2 Another sub-branch of "Third Way Thinking" is evolutionary economic geography (see the special issue of Economic Geography, (2) 2009) and with key advocates such as Boschma (2004), Hodgson (2005), Grabher (2009), Boschma and Frenken (2006), among many others. I will not discuss in detail this sub-branch as it has many overlapping with NR, and it is more theoretical than policy oriented. For a thorough critique see Mackinnon et al. (2009).

3 Key thinkers include Krugman (1991), Fujita and Krugman (2004) and Venables (1996). To be fair, Krugman in his New York Times column has been repeatedly very critical of austerity imposed on Southern Europe and of the Troika's policies. Here I refer to Krugman's geographical/regional contributions and how they have been used by others.

4 This section draws heavily from Hadjimichalis (2006).

5 I will not discuss here the major issue of migrants and refugees "without papers" in the EU, the invisible 29th state, a map you cannot draw, of approximately 10–13 million people (see also Chapter 2). They live, work and produce wealth with or without limited social rights, not being subjects of justice to the same degree as other EU citizens. Among the extensive documentation, see Anthias and Lazaridis (2000), EC (2007), Dijstelbloem and Meijer (2011), Mezzadra and Neilson (2013), Lafazani (2014) and Trikliminiotis et al. (2015).

6 The first and most important consequence of the Greek crisis was the euro's fall by 12 per cent in only 3.5 months in early 2010, a true kiss of life for the German economy: New orders for its industrial sector increased considerably bringing about the greatest climax since 2007. In the entire Eurozone, the economic situation improved for the very first time after the recession so that the then French finance minister Christine Lagarde argued "Europe became more competitive because of Greece". And even in Greece, in the middle of its public debt crisis, industrial capital gained considerably with a 25 per cent increase in exports during 2010 relative to 2007.

7 For an important re-introduction of questions on spatial justice, see the bilingual electronic journal Justice Spatiale/Spatial Justice, www.jssj.com.

8 The Social Justice Index in this study is composed of six dimensions: poverty prevention, equitable education, labour market access, social cohesion and non-discrimination, health and intergenerational justice. I have major reservations about the chosen dimensions and the method used but since, to my knowledge, there is no other available study I use it with reservations.
9 Severe material deprivation exists if persons cannot afford at least four of the following items: (1) to pay rent or utility bills, (2) to keep the home adequately warm, (3) to face unexpected expenses, (4) to eat meat, fish or protein every second day, (5) a week's holiday away from home, (6) a car, (7) a washing machine, (8) a colour TV, or (9) a telephone.
10 "At risk of poverty" is defined as those persons with an income equivalent to 60% of the national median equivalized with disposable income after social transfers.
11 Percentage of population 18–24 with at most lower secondary education and who are not in further education or training.

6 "Nobody alone in the crisis"[1]

Resistance and solidarity

...I don't believe in charity; I believe in solidarity. Charity is vertical, so it's humiliating. It goes from top to bottom. Solidarity is horizontal. It respects the other and learns from the other. I have a lot to learn from other people.

Eduardo Galeano (2004: 146)

...The movements become movements by occupying space

(Castells, 2016)

Every day since January 2012, dozens of people of all ages from all over Athens come for medical treatment in an unusual building in the former US military air base at Hellinikon, a neighbourhood in Southeast Athens. Two hundred volunteers work in shifts (doctors, nurses, technicians and administrative staff) to provide free health and pharmaceutical services to people in need, Greeks and migrants alike. During the first five years of its operation, the Metropolitan Social Clinic at Helliniko has treated more than 35,000 patients, from simple paediatric cases to difficult cancer treatment and chemotherapy. By the end of 2013, unemployment in Greece was at an estimated record high level of 29 per cent. Approximately 1.4 million people, about half of them long-term unemployed, had lost their health insurance after losing their jobs through the introduction of an inhuman law in 2012. As austerity measures continue, and the so-called "recovery plan" continues to fail, with declining GDP for six successive years, matters are only getting worse. And nowhere else is this more evident than in the destruction of the national health system. So self-organised social clinics, like the one at Hellinikon, are the only solution for thousands of poor and uninsured people. This extreme situation has mobilised an extraordinary solidarity network of social clinics and pharmacies all over Athens and the rest of Greece. By 2016, more than 50 similar social clinics and pharmacies were in operation across the country, mainly in major cities, based on volunteer work and donations only of equipment, drugs and other materials. No money donations are allowed, no political parties or other organised groups are allowed and no one gets publicity for their donation. Dr Giorgos Vichas, a cardiologist, says in a video on the clinic's website

"...the kind of solidarity which is built here every day is the b asis for the society we envisage for tomorrow".

On 15 May 2011 in Spain, the week before regional and municipal elections, the digital platform "Real Democracy Now" (*Democracia Real Ya, DRY*) called on "the unemployed, the poorly paid, the precarious, the young people, the homeless...." to take to the Spanish streets. Despite the boycott by all mainstream media, hundreds of thousands of people took to the streets in 50 Spanish cities, informed by social media. The 15M mobilisation, as it became known, quickly gave way to the occupation of major squares in most cities, the largest being at (PSM) Puerta del Sol in Madrid and Plaza Cataluña in Barcelona. *Indignados*, as they call themselves, stayed at these squares until mid-July, despite elections and the threat of police intervention. It was the peak of the Spanish resistance movement and a prototype for immediate democracy and self-organisation that influenced other square occupations around the world. Participants organised several committees, working groups and assemblies to deal with day-to-day issues such as cleaning, nutrition, communication, action/performances etc. What happened after the end of the camp at PSM is described by Silvia Nanclares and Patricia Horrillo (2017: 22) as "...the transformation of the spirit of 15M into a sort of virus. The virus was spreading despite the system's defense mechanisms".

These are but two among hundreds of social movements across SE during this era of crisis, some protesting, as in Puerta del Sol, while others provide solidarity assistance to those in need, as in Hellinikon. All, however, are based on and reproduce spatialised democratic politics through dense social networks, local as well global, via face-to-face and digital communication. Despite major differences in terms of target, organisation and methods, they have two common characteristics. First, they highlight the fact that *many people in SE are active agents in resistance and solidarity social movements – often with a radical left core, not accepting to be passive victims* of the crude ultra-neoliberal policies applied in their countries. Second, they highlight that *the prime field of struggle, in particular urban socio-spatial settings, is everyday life. They demand social, spatial and environmental justice and defend democracy and dignity in public space, both materially and symbolically.*

For SE, the major structural factor behind these mobilisations and solidarity actions is ultra-austerity policies, the extreme material deprivation of everyday life, political corruption and the absence of democratic accountability that together produce anger and the strong feeling of injustice described in previous chapters. In this one, I would like to focus on the role of social agency in understanding who has the capacity, with or without experience from other movements, to mobilise during crisis and how, while having the courage to resist austerity and to build networks of solidarity and social movement structures. This raises questions about what shapes and limits different popular crisis responses, not only for resistance and solidarity, but also for re-orientation and the emancipation of social agencies.

These questions re-open the famous Leninist/Maoist dialectic between objective and subjective conditions together with the Gramscian demand for alliances and solidarity across social divisions and spaces. In other words, we need to distinguish between the everyday mechanisms through which people simply "get by" and those activities that allow people in social movements to create alternative spaces of resistance and survival where political emancipation can take place. Movements in the period from 2010–2015 in SE took place at different scales. Some were local and independent while others were networked nationally and internationally, building strong political and social alliances. Three preliminary remarks are necessary:

First, the social movements described below bear the contradictory characteristic of being highly uneven and internally diverse but quite similar at the same time. On the one hand, they are responses – but not in causal relationships – to concrete applications of "real existing" neoliberal austerity that takes uneven forms and intensities in every SE social formation. Different militant particularisms, to recall Raymond Williams, embodied in place produced uneven actions and outcomes. On the other hand, common activists' experience across Europe against the undemocratic EU and Eurozone regime, the ability to communicate and transfer tactics from place to place and to build international solidarity networks resulted in similar demands, actions and spatialised politics.

Second, while I acknowledge particular historical and spatial conjunctures in SE during the crisis, plus the current technological ability for digital communication, I want to highlight that these forms of action have surfaced and resurfaced many times in the past and that the context of economic crises and hard austerity give them particular frequency and intensity. Social movements reproduce traditions of protest and solidarity that live in the collective memory of societies. As Della Porta (2015) argues, social movement activists reproduce similar forms of action over time because they are what people know. And Featherstone (2012) describes the roots of social struggles and solidarity in early labour disputes and anti-colonialism and via Gramsci focuses on more recent anti-global events in Seattle, the (WSF) World Social Forum and the (GJM) Global Justice Movement. Thus, the tendency of some researchers and the mainstream media to describe these movements as very new – in order to promote the novelty of their ideas – succumbs to the narratives of the Short Past and ignores the years-long SE rebellious tradition of organising and fighting in public spaces for social demands.

Third, despite extensive documentation on the existence of different anti-austerity mobilisation forms in SE, the majority of the literature focuses, almost exclusively, on the squares' movements while giving scant attention to massive union-led demonstrations and to solidarity movements (for exceptions see: Della Porta, 2015; Baumgarten, 2013; Diani and Kousis, 2014; Vaiou and Kalandides, 2016, 2017; Papataxiarchis, 2016c). General strikes plus rallies by political parties, students and anarchist groups and the work of hundreds of (NGOs) Non-Governmental Organisations and bottom-up, self-organised solidarity initiatives remained largely unnoticed or undervalued in terms of their importance. I don't deny that some unions were and remain corrupt, or associated with political parties and practices that had a large measure of responsibility for the present situation in SE, while NGOs had and have problematic relations with the power establishment on which they rely for funding. However, instead of addressing these problems, the geographic and urban/regional studies literature simply ignores the unions and the solidarity movements, thus giving an incomplete or wrong picture. In my analysis, I distinguish four forms of resistance and solidarity actions. First: union-led demonstrations and strikes in which political parties, student organisations and others participate; second: independent small-scale protests and social movement networks; third: occupation of public squares and other public/private buildings; and fourth: informal self-organised solidarity social movements outside the institutional structures of the state, municipalities, the Church and NGOs (see also LIVEWHAT, 2016; www.solidarity4all.gr; Rakopoulos, 2016; Vaiou and Kalandides, 2017; Zavos et al., 2017).

Political economy has reflected on several aspects that are important for our understanding of social movements. In particular, it has helped in delineating three spatio-temporalities of resistance to capitalist austerity and social survival under it.

Firstly, a long-term spatio-temporality refers to general regimes of capitalism and its reproduction through periodic crises, as in the case of rentier capitalism. Rent-extraction activities have broad and profound social effects on the ways people react and mobilise. Due to declining productive activities in SE, popular struggles along the capital-labour conflict in production also articulate themselves around issues of property, privatisation, welfare services and the environment. They concern the distribution of the value created by rents, and they bring together heterogeneous publics politicised along issues of social reproduction and socio-spatial justice. We need to look, as Fraser (2013) proposes, to the "...broader terrain of *social reproduction*, not only to class relations ... as the sole or principal ground of political struggle" (p. 125, emphasis in the original).

Moreover, a middle-range spatio-temporality singles out cyclical shifts of growth and crises in particular social formations, as in the case of the current crisis in SE. Massive and spatially uneven unemployment and deprivation of social reproduction produces socio-spatial inequalities and injustices. Socio-spatial injustice is a powerful symbolic force that works effectively across class, race, gender and place. In this respect, it may create collective consciousness and strong solidarity ties when particular groups of people believe they are no longer equals in political terms and mobilise to protest injustices or to build solidarity organisations helping others.

Finally, there is a contingent spatio-temporality that affects socially and spatially uneven classes, different ages, genders, ethnic groups and migrants in particular places. The urban emerges as the prime, although not the only, field/place in which the contingent spatio-temporality reproduces different forms of protest, resistance and solidarity. These three spatio-temporalities operate simultaneously in a multi-scalar framework. They are dependent on the interplay of endogenous and exogenous factors, and finally, they form particular struggle regimes that fuel both peaceful and violent upheavals, namely the ensemble of uneven social and cultural practices used by activists under concrete constraints imposed by the dominant powers.

Keeping the traditions of socio-spatial struggles alive

The streets and squares have been returned to politics in SE, but with new demands and tactics.[2] As many commentators have observed, from 2008 until 2014–2015, a persistent sequence of urban protests and insurrections took place in SE and all over the world. Solidarity networks and protest against austerity across SE during the crisis period did not arise unexpectedly. On the one hand, the degradation of everyday life, unemployment, homelessness, precarity and anger against corrupt politicians is more intense in urban areas (Benach, 2015). Neoliberalism and austerity disproportionately target urban areas in SE, as they have done in the rest of the world. "Austerity bites" in cities, as Peck (2012) has pointed out. It is therefore not surprising that social resistance acquired particular intensity in cities,[3] as has always been the case. Besides unemployment and the collapse of social and technical infrastructures, cities are strong bastions for tertiary sector unions, for left-wing parties and for smaller radical and anarchist groups and are also an important locus of student radicalism. In addition, rentier capitalism and the crisis of real estate and housing made cities in SE the obvious protest arenas, although in Greek cities, the collapse of the housing market was not as severe as in the other countries. With the exception of social movements related to housing evictions and occupied social centres, other forms of struggle did not arise from urban specific contradictions. Although they took place in cities re-politicising public space, are not urban social movements as such. As Margit Mayer (2013) demonstrated, urban mobilisations under neoliberalism are highly diverse internally, accommodating actors from militant activists to middle-class lifestyle seekers, and their impact is contingent to place/time particularity.

On the other, a lived but not path-dependent tradition exists in the four countries connecting struggles during the crisis with other struggles dating back to the 1970s and 1980s. Older struggles belong to different families of protest, depending on the leading force, the target and the cause. Trade unions and political parties organised some of them, others were youth-led and were spontaneous and rebellious and yet others were local/national or transnational. All, however, took place in southern cities, or in other European cities if they were transnational, with the participation of Southern activists. These events point to the fact that the urban scene, rather than places of production, was the prime field of social struggles and resistance that demanded democracy and the end of austerity. This distinction, however, does not mean that the capital-labour relation was absent from urban protest – quite the contrary. Protesting against the imposed policies on wage cuts, public servant layoffs, increased taxation and changes in labour laws, among others, was a direct confrontation by labour against domestic and EU austerity policies favouring capital.

Resistance in SE during the 2000s – with actions that had more dissimilarities than similarities across regions and cities – is distinguished by the *more comprehensive political use of "the urban" and particularly the city centre as a major field of struggle.* By political use of the urban, I mean not only the necessary bodily presence of protesters in streets and squares, but also three other important parameters.

First, the past and crisis-led experience from older urban social movements arose from contradictions generated in the secondary circuit of capital in the built environment. Activists acquired experience by attempting to achieve control of their urban environment with actions such as preventing evictions, squatting empty buildings, fighting against gentrification, protecting green areas, etc. (for more see Koutrolikou and Spanou, 2013; Mayer, 2013; Siatitsa, 2014). These activists exerted great influence on the 2011 SE upheavals.

Second, the active involvement of left-wingers and anarchists as individuals in these struggles, particularly in Italy, Greece and Portugal, draws from older urban militant traditions and extended the then current ones into anti-austerity and anti-capitalist issues. It may well be that some of the celebrated urban events during 2011 were spontaneously initiated by young people using digital communication, but their post-spontaneous organisation and management, with committees, assemblies, working groups, tactics against the riot police etc. was made possible by the presence and involvement of experienced activists.

Third, there was the organisation of bottom-up, informal solidarity initiatives engaged in a large repertoire of actions at different scales with diverse targets and actors, from social kitchens, time banks and alternative currencies to food distribution, housing, culture and healthcare (Cabot, 2016; Vaiou and Kalandides, 2017). Organised periodically in streets and squares, or operating in stable premises up to today (2017), they constitute an attempt to cope with and also resist the everyday dramatic degradation of people's lives and the gradual loss of their dignity. Let me clarify these points with a short narrative of past experiences in this section and with a discussion of post-2010 resistance movements and solidarity initiatives in the next two sections.

Starting in the 1970s, spatial organisation and planning became highly political during dictatorships. Social demands for improving quality of life at the local scale were considered "legitimate" by authoritarian governments, and several urban, regional and environmental movements appeared in Spain, Greece and, under a totally different political regime, in Italy.

The case of urban social movements in Madrid (1970–1979), described by Castells (1983: 217–224), is paradigmatic of how urban movements grew and kept connections with the goals of democracy (and sometimes socialism) and how the underground left parties found opportunities to reach people while remaining less exposed to Franco's police. Many left-wing activists took part in these movements as preferential entry points to contact and emancipate people. In Greece, similar developments were observed in environmental movements, mainly mobilisations against polluting industries and cases of urban squatter evictions (Hadjimichalis, 1987). In this respect, uneven spatial development patterns at multiple scales acquired a strong political dimension, and in the 1970s, after the fall of the dictatorships, many leaders of these urban/regional/environmental mobilisations became left-wing delegates in unions, political parties and city councils. A case in point is the left-wing coalition led by PSOE that won the municipal elections in Barcelona in 1979. It was based on collaboration between social-movement activists active since the late-Franco period, linked to neighbourhood associations (Eizaguire et al., 2017).

In Italy in the 1970s and 1980s, two different urban social movements took place. On the one hand, there was the "mainstream" left urban regeneration movement with the paradigmatic case of "Red" Bologna, a historical bastion of PCI, where left-wing planners and architects applied themselves to the participatory redevelopment of the historic centre while avoiding gentrification (Cervellati and Scannavini, 1973). On the other hand, we have the more radical left experience of squatting and occupying self-managed social centres *(Centri Sociali Occupati Autogestiti)*. They were used as meeting places, housing and cultural centres and providers of various social services to marginal groups and to near neighbourhoods (see Box 6.1).

Box 6.1 Occupied social centres in Italy

The (ISCs) Italian Social Centres appeared in the late 1970s and 1980s in large abandoned spaces and buildings in the periphery of large cities, usually former factories, military camps and schools. They became known as *Centri Sociali Occupati Autogestiti*, where new forms of social, cultural and political action took place, beyond the market ideology, profit and norms of the state and without exchange in money form. ISCs provided a variety of services including shelter, day-care services, education, health clinics and the exchange of food and clothes and were clearly anchored in local neighbourhoods. They had historical relations with the non-parliamentary left, particularly Autonomia Operaia and Lotta Continua and with various anarchist groups. Clashes with the police were frequent but massive participation and organised resistance helped them to continue. People from ISC took part in all massive demonstrations and alternative social movements across Italy including the Tute Bianche (people in white working overalls) and Ya Basta (Enough!) in the 2000s during the Global Justice Demonstrations across Europe. Among those ISCs with long existence and socio-political importance are Leoncavallo in Milan, Spartaco in Ravenna, Pendro in Padova, Corto Circuito and Triburtina in Rome and Oficina 99 in Napoli. Relatively new ones include Insurgencia in Napoli, Forte Prenestino in Rome and Askatasuna in Turin. A process of legalisation from local authorities started in 1993, and by 1998, 50 per cent of ISCs had legal status – which generated bitter ideological fights within the movement. The Italian ISCs inspired many similar movements in Spain but very few in Portugal and Greece.

Sources: Cecchi et al., 1978; Vaiou and Hadjimichalis, 2012; Siatitsa, 2014.

In the 1970s and 1980s, the Gramscian tradition was strong in Eurocommunist circles, and this permitted, in Italy, Spain and Greece,[4] the hesitant introduction of urban issues and questions of everyday life in the party's agenda. We cannot speak of an urban political consciousness per se at that time, but can characterise this as a first attempt to incorporate questions of social reproduction at the local, neighbourhood scale into the dominant worker-based ideology of the parties. The latter was applied via a double membership: first, membership/participation in unions or associations related to one's work (a sector-based political enrolment/attachment); second, and at the same time, participation in local neighbourhood committees related to one's community (a spatial-political enrolment/attachment) that followed in the path of urban social movements during dictatorships. This double political enrolment was not far from what Antonio Gramsci proposed back in 1919 as "neighborhood Circles", in *L' Ordine Nuovo*:

> …A vast field for concrete revolutionary propaganda would open for communists organized within the Party and within neighborhood Circles. These Circles, in collaboration with the urban Sections (of the Party), should make a census of the working classes (living) in the area and become the seat for a neighborhood workers Council of delegates, the link that brings together and concentrates all proletarian energies in the neighborhood …

Gramsci goes on to explain how these local "Circles" should evolve into dynamic nuclei in each neighbourhood and how elected representatives, one for every 15 workers "…as in English factories", should represent all categories of working persons, and not only industrial proletarians. And he adds:

> The neighborhood Council should be ready to incorporate delegates from other categories of workers who live in the area: waiters, coachmen, scavengers, railway employees, tram drivers, private white collar workers, shop assistants etc. The neighborhood Council should emanate from all working classes living locally, a legitimate and authoritative representation, able to respect discipline, invested in spontaneously delegated power, able to coordinate an immediate and full stoppage of any work in the neighborhood. Such a system of proletarian democracy (integrated with equivalent organizations of the peasantry), would give form and discipline to the masses, becoming a magnificent school of political and administrative experience...
>
> ("Democrazia operaia", *L' Ordine Nuovo*, 21 June 1919, 1995 edition; emphasis in the original, my translation from Italian)

Although PCI ruled major cities in Italy such as Firenze, Perugia, Ancona and particularly Bologna, with many positive – at that time – outcomes, the Gramscian proposal was only partially applied. In the 1970s and 1980s the PCI continued to prioritise factories and unions as fields of struggle, and this permitted non-parliamentary left parties, such as Lotta Continua, Potere Operaio, Sinistra Proletaria and others, to focus on urban social issues and everyday life as legitimate foci for (class) struggle, under the slogan "Let's take the City" (*Prendiamoci la Città*) (Siatitsa, 2014). In Greece and Spain the Gramscian tradition remained mainly an intention by isolated left-wing activists,[5] while in Greece an anarchist spatial tradition similar to Spain's never developed.

Dimitra Siatitsa (2014) argues, based on extensive research, that in Spain and Italy during the 1980s and mid-1990s, and decreasing in intensity until the mid-2000s, urban social movements focused mainly on the housing question. Despite real estate booms, thousands of young people, poor families and migrants were excluded from access to affordable housing. In both countries, dozens of occupations took place, and local young activists organised in "fighting committees" such as "Okupa" in Madrid and Barcelona. These activists rejected collaboration with other local actors, such as the neighbourhood associations. In Italy, after 2005, housing movements acquired new strength and formed networks such as the *Comitato di Agitazione Borgata* and *Comitato di Lotta per la Casa* in Rome. Fights with police to prevent evictions were frequent.

Along with major changes in the SE economies and societies, and the wider defeat of radical left ideas after 1989, the communist and socialist left has been gradually discredited due to corruption and scandals in left-led local administrations, particularly in Italy. Similar problems faced socialist parties in central government in all four counties, and so did the left-wing identity in urban and regional struggles. New autonomous political subjects appeared, and new demands came to the fore, not necessarily on the left, best illustrated by the anti-global, anti-war and particularly the ecological movements and the GJM. It was the period of transnational radical activism against G8, the (WTO) World Trade Organisation and the imperialist intervention in Iraq, with SE activists participating in all major mobilisations in Prague, Genoa,[6] Cannes, Brussels, Thessaloniki, Malmö, Rostock and other cities. In addition, the 1990s union-led European Marches targeting EU summits in Amsterdam, Cologne, Nice and Gothenburg attracted thousands of protestors across Europe, SE included. The arrogance and brutality of the elites was evident in most of the transnational demonstrations when activists were refused entry at the borders, preventively arrested or attacked by aggressive riot police forces.

In these transnational, urban mobilisations, young activists encountered older ones and learned about international solidarity, about past experience in movements and, not least, about tactics against riot police forces. In exchange, older activists learned from the young generation the ecological discourse, the achievements of social economics and the effective use of the internet and social media for networking purposes. Above all, both developed a consciousness of "belonging" to an international radical multitude, some with a direct connection to anarchism, others to left-wing parties, yet others to independent environmental and feminist groups, etc. For Greek activists, the combination of the anti-Olympic 2004[7] and anti-global/anti-war movements had, for the first time since 1989, a major political effect: a sense of again being a part of an international movement, which fights the class enemy on different fronts and spatial scales.

Along with the transnational ones, social movements against the elimination of civil rights, racism, fascism, police brutality and exploitative working conditions forced on thousands of economic migrants, with or without papers, since the 1980s, played a major role in shaping radical political identities in SE. Among the first collectives/groups that opened this agenda was the Network of Social Support to Migrants and Refugees, established in 1986 in Athens and operated on a voluntary basis without any support from the state or NGOs. Known as *To Diktio* (The Network), it runs an occupied social centre for migrants in the centre of Athens and organises every year a three-day anti-racist festival. Daily activities include free Greek language and computer courses, free legal advice and active solidarity in several situations, such as supporting migrants detained in police stations and Detention Centres, providing support in health, housing and labour issues, etc. (Lafazani, 2014). Similar movements and social centres exist in Spain in Malaga and in Italy in Naples, Bologna and Milan. Most of those centres and movements are part of the "No Border" network, organising camps, activism and innovative documentation of repression against migrants and migrant deaths at European borders using radical cartography (see www.noborder.org).

In this respect, the radical tradition persisted in SE, this time along with international activism, combining social and spatial struggles. A crucial turning point, following the (WSF) World Social Forum in Porto Alegre, was the birth of the (ESF) European Social Forum, which, in its first meeting in Florence in 2002, adopted a "Charter for the Right to the City" following an Italian/Spanish/Greek proposal (Portaliou, 2007; Siatitsa, 2014). The Florence ESF was followed by Paris and London and finally by Athens in 2006. All forums played a major educative and communicative role in SE social movements. In these meetings, during discussions with other European movements that had more experience because of their longer existence and action away from political parties, SE activists learned procedures of direct democracy and built up an attitude towards social, multi-class mobilisation and actions, parallel, but not conflicting with, traditional class actions. In the Athens ESF, there was a large special section featuring informational kiosks and meetings hosting urban and environmental social movements, housing struggles, the application of the right to the city initiative in different European cities, a participatory budget, movements against the privatisation/dispossession of public utilities, social identity movements and alternative, radical forms of municipal government.[8] It was a clear attempt to put class issues at the point of production together with gender and race issues, including wider socio-spatial ones related to the field of reproduction and everyday life at different geographical scales.

In ESFs, younger generations of activists met older ones from the 1980–2000 period and together sowed the seeds for a new kind of radical "we". This, in turn, made possible its political appearance later during the crisis thanks also to the technological innovation of digital communication. Thus, during the 2000s, three parallel processes took place in SE societies. First, the crisis produced new marginal political subjects, highlighting the importance of everyday life and social reproduction and pushing to the forefront demands for housing, welfare services, the environment, dignity and democracy. In short, it re-introduced the "right to the city" with demands that could mobilise everyone affected. Second, at the social-political level, a steady increase of involvement by young people could be seen, the "precariat", in radical, social and urban struggles, not associated with unions and without clear political party affiliation. Older generation activists, usually on the left, were there but they didn't set the tone. Third, at the communication level, the coming together of individuals and social groups via digital communication and the organisation of dozens of digital platforms to spread information was crucial to but not the reason for the upheavals.

"They want us precarious – We will be rebellious!"[9]

A wave of social mobilisations, known as the square's movements, exploded in 2011, echoing for some analysts the events of 1968 with similar massive demonstrations and occupations. The names of several squares became symbols of resistance around the world, such as Syntagma, Puerta del Sol, Rossio, Taksim, Tahrir, Zuccotti Park and many more (Ancelovici, et al., 2016). These protest movements emerged within a specific time frame, following what has been called the "Arab Spring", and occupied central, symbolic squares in each city. Their simultaneity and the spatial reference to squares is not, however, a sufficient reason to put them in the same category and to draw general conclusions. They belong to different families of protest, with unique cultural features and political demands. In what follows, I focus on protest events in SE only that reacted to injustices and undemocratic politics within the Eurozone family, underlining their diversity and their uneven development and impact.

Assembled for different reasons but under a common demand for Real Democracy Now and using similar digital methods of communication, these movements quickly universalised their claims to ask for and imagine a real restructuring of political life against austerity. This was a moment of real rupture with the depoliticising practices that had marked mainstream politics in previous decades. These events, as argued by Wilson and Swyngedouw (2015), challenged arguments for de-politicisation in post-politics, as the main characteristic of our times, where democratic politics are replaced by technocratic mechanisms, free markets and liberal cosmopolitanism. In post-politics, there is no history, no social classes, no socio-spatial inequalities and no injustice. Politics is reduced to the consensual management of economic necessity. The events in SE, and globally, proved the opposite via the spatialisation of democratic politics. The 15M manifesto from Puerta del Sol makes the point:

> We recovered and utilize the public space: we occupied the squares and the streets of our cities to meet and work in a collective, open and visible way. We inform and invite every citizen to participate. We debate about problems, look for solutions and organize actions and mobilisations. Our digital networks and tools are open: all information is available on the Internet, in the streets and in the squares (www.takethesquare.net/es/2011/08/18/como-cocinar-una-revolution-no-violenta/)
>
> (From the 15M manifesto, Puerta del Sol, Madrid, 2011)

The first and most frequent form of anti-austerity protest in SE, however, since February 2010 when the crisis hit first Greece, was massive union-led street demonstrations, rallies and strikes. Square movements attracted the attention of the international press and several researchers, but they were by no means the only or most important resistance struggle in SE.

Union-led rallies were either sector-based (e.g. teachers, taxi drivers, civil employees, pensioners, private and public sector workers and others) or general strikes including all economic sectors, modes of transport, air traffic controllers, even police officers, coast guards and firemen. Among SE countries, Greece stands out for both the severity of austerity measures on people's lives and for the extent of mobilisations. Diani and Kousis (2014) quote Greek police records that show that in the period from March 2010–March 2014, there were over 20,000 contentious episodes across the country, of which 31 were large (over 5,000 people involved). From these large protest events, 11 were non-union led (students, political parties and Syntagma square occupation) and 20 were union-led, all organised in urban areas. Initial general strikes attracted several hundred thousand people to the streets when the majority of MPs in parliament voted for austerity measures. Rallies ended frequently in quite violent battles between militant protesters, the SE version of the Black Block, and police, with protesters throwing stones and Molotov cocktails while special riot police forces reacted with tear gas, clubs, punches and kicks (see box 6.2).

Box 6.2 A black day in the anti-austerity struggle

On 5 May 2010, in response to the proposed spending cuts, a nationwide strike was called in Greece. Starting at midnight, air traffic, trains and ferries ceased. Schools, some hospitals, public services and many private businesses were closed. An estimated 100 to 300,000 people marched in Athens, and when some protestors tried to storm the parliament, riot police pushed the crowds back and a real battle started with tear gas and flash bombs from the police and Molotov cocktails from protestors. A group of ultra-leftist protestors of the Black Block set fire to a small private bank while people worked inside, ignoring the general strike over fears of losing their jobs. Three people died that day, among which was a pregnant woman, despite efforts by fire crews and other demonstrators – black day indeed in the Greek anti-austerity struggle.

In Spain, a local rally was organised in Barcelona in January 2010 against the EU summit (*Contra l'Europa del capital, la crisis i la Guerra*) by the anarchist union CGT, radical students and human rights organisations. And in June 2010, CCOO and UGT (*Unión General de Trabajadores*) unions called a national rally against labour reforms. A major reaction took place on 29 September 2010 with the organisation of a national general strike, the first after 2002, as part of the Pan-European Trade Union Confederation action. The strike marked a break in the once-close relationship between unions and the socialist government. Tens of thousands of demonstrators poured into Brussels to join a 100,000-strong march on EU institutions and reinforce the impact of general strikes in Spain, Greece and Portugal on the same day (www.nbcnews.com). Greek doctors and railway employees walked out, Spanish workers shut down trains and buses, Portuguese workers blocked ships at ports and flights at airports while Italian metal workers walked out from factories. Innovative demonstrations, known as "colourful tides", have been organised since 2013 in Spanish cities, with people wearing different-coloured working overalls (an influence from the Italian Tute Bianche) to protest against austerity in particular sectors: white for public health, green for public education, blue against water privatisation, purple for women's rights and black for public servants (see www.huffingtonpost.es). All such actions sought to protest the budget-slashing, tax-hiking, pension-cutting austerity policies of their governments, some policies seeking to control debt, others to escape from a bailout possibility. General strikes are relatively rare in Europe, but from 2008 until 2013, Greece had 16, Portugal six and Spain four.

In Portugal, according to Accornero and Ramos Pinto (2015), contrary to the emphasis on "new new" social movements, labour has remained throughout as the most significant protest actor, even if it has been forced to look for new allies. Across this period, 78 major protest events were initiated by public-sector unions and 11 were organised by the national trade union federations (CGTP or UGT). Overall, two-thirds of protests arose from the labour arena and workers' representative organisations. In contrast, new social movement organisations such as QLT, M12M or Geração à Rasca appeared as the lead actors in only 19 protest events.

The first coordinated trade union action against austerity took place in Spain, Portugal, Italy and Greece in November 2012. Organisers claimed that half a million protested in Barcelona and 350,000 in Madrid, although police gave lower figures. In Lisbon, there were around 20,000, but clashes with police left nearly 50 hurt, and in Madrid, 70 were injured and some 40 arrests were made (www.bbc.com). In Athens, participation was low with 10,000 protesting, while in Thessaloniki dozens of protestors broke into a conference centre to protest the presence of German officials in a meeting with mayors. In Turin and Rome, participation was also low, 15–30,000 people, but with violent battles between protestors and the police.

Against this union-led, militant and highly politicised protest background that followed a more traditional use of urban space, as well as top-down political organisation and defensive demands, another form of resistance took place. This was the occupation of the capital's symbolic square in each country followed by similar occupations in other SE cities.

Spain, Puerta del Sol and Plaza Cataluña

In (PSM) Puerta del Sol, the actual revolt started after the first camp was broken up by the police on 16 May 2011 after only one day. The eviction was spread on social networks, and more than 5,000 went the same evening to camp at PSM. Protest against the police found support from new occupations in Barcelona, Valencia, Seville and many other cities (Sevilla-Buitrago, 2015a). According to Carlos Taibo (2013), the *Indignados* belonged to two different protest groups. First, there were activists who had been participating quietly and steadily for several years in different social movements before 2011, including self-managed and occupied social centres, feminist groups, environmental groups and networks of solidarity. Similar observations were made by Conill et al. (2012: 244–245) who argued for the Plaza Cataluña occupation:

> ...many of those who are...participants in alternative economic practices were present in the "indignants" movement. For them, there was a logical continuity between their distance from the norms and institutions of capitalism and the protest against the indignity of political leaders that, in their view, led most people in Catalonia, Spain and Europe to the dead end of the crisis in the labour market and in social services, while banks recorded unprecedented high levels of profits (....) Thus, the alternative economic culture that preceded the economic crisis, by virtue of its present critique of capitalism, rose to the forefront of the public debate.

Furthermore, activists from Okupas, the housing and social centre occupations movement, was the first group who decided to camp and sleep at the PSM. The protest camp that subsequently formed and occupied the square for nearly a month followed the self-management methodology of the squatters' social centres.

Finally, the M15 was made up of young people who mobilised for the first time. Visible deterioration of their everyday life at three levels had made them angry and rebellious: universities where scholarships had been replaced with credits; the labour market with record-high unemployment and *contratos-basura* (rubbish contracts); and the housing market where they had no access to housing, such that 70 per cent of 18–30-year-olds were living with parents or grandparents. The slogans in Puerta del Sol by those young people, "Youth without Future" (*Juventud sin Futuro*), "Homeless, Jobless, Pensionless, Fearless" (*Sin casa, sin curro, sin pension, sin miedo*) appeared everywhere (Observatorio Metropolitano, 2013).

Slogans and placards in PSM were about political corruption, capitalist greed, precariousness and homelessness. It was a framing that characterised the actions and policies of a national power system (particularly the government, from national to regional and municipal) as unjust and corrupt, both providing legitimate grounds for indignation. As the 15M manifesto "How to Cook a Non-Violent Revolution" (Cómo cocinar una revolución no violenta) said, "We don't understand why we need to pay the bills of a crisis whose authors continue to enjoy record benefits. We are fed up with injustices". Present in the demonstrations were also left-wing people in their 60s and 70s and others who faced severe cuts in pensions and health security. PMS was occupied during the national elections in which the right-wing PP (Partito Popular) gained the majority while PSOE faced a historical defeat, victim of its neoliberal policies. The camp, however, continued to exist for another three weeks, doubled in size and developed better and more sophisticated organisation, with the general assembly playing the major role (Observatorio Metropolitano, 2013; Sevilla-Buitrago, 2015a). After the global action day of 15 October 2011 with half a million protesters in Madrid, several thousand activists of the 15M movement decided to occupy a closed space, the Hotel Madrid, an abandoned hotel close to PDS (Abellan, Sequera and Janoschka, 2012). A group of *Indignados* and experienced squatters took the lead to establish a visible "home base" and to accommodate evicted families and other vulnerable groups. On 5 December 2011, the police evicted the squatters but "...the hotel still remains as a powerful symbol in the movements' collective memory as a laboratory for urban resistance" (p. 321)

Portugal, Rossio

As noted in the book edited by Mark Bergfeld (2014) *Portugal, 40 years after the Revolution*, while the international press paid attention to the *Indignados'* protest in Spain and the square occupations in Greece, the revival of Portugal's social movements went unnoticed. In fact, the Portuguese protest of the *Geração à Rasca* (The Desperate Generation) on 12 March 2011, which took place two months before the Spanish *Indignados'* action, was the biggest demonstration in Portugal since the Carnation Revolution of 1974, with 400–500 thousand participants in the streets, mainly young people (Baumgarten, 2013). The mobilisation was also the starting point of important changes in the content and organisation of Portuguese protests against austerity, the Troika and the government. A song, "Parva que sou" (How stupid I am), performed at a huge concert in Lisbon in January 2011 by the band Deolinda, inspired the demonstration. This song is about the precarious situation of a generation of educated people and initiated the idea for a digital call for the 12 March 2011 protest.

Many activists consider Portuguese people as less involved in civil society groups or alternative social movements and less politically active, despite the fact that major demonstrations took place in the past. Among them, I recall, in 1983 against the IMF, against joining the EU in 1986, mobilisations in the context of GJM and the ESFs in 1999 and against the Iraq War in 2003. Groups remain small, and activists complain about the lack of support away from Lisbon and Porto. Baumgarten (2013) distinguishes three mobilisation categories in Portugal after March 2011. First, major union-led demonstrations and general strikes, which became more frequent after 2007: four during 1982–2002 and six from 2007 to 2013. Second, independent protest events and social movement platforms that began in 2011. The activist platform 15.0 (October), formed during the occupation of Rossio Square in Lisbon, called for protest on different occasions, some of which were part of an international demonstration in Lisbon and Oporto with thousands of participants. And third, occupation of public spaces, beginning on 20 May 2011, as spontaneous actions by small groups, inspired by the Spanish M15 movement.

The largest was in Rossio Square, where more than 100 people occupied the square for three weeks and camped overnight. As Luhuna Carvalho (2017) describes, Rossio's occupation didn't build upon previous networks of social, countercultural, or political movements as happened in Puerta del Sol and Plaza Cataluña. She also points to the particularities of Lisbon's centre, which is practically empty of young people as residents, the opposite situation from Athens. In this respect, the Rossio movement was an important rediscovery of the city centre as political territory.

According to Accornero and Ramos Pinto (2015), towards the beginning of 2012, anti-austerity movements split along the traditional divisions of the left between the (PCP) Portuguese Communist Party and the Left Bloc (*Bloco de Esquerda*). However, the establishment of the Democratic Congress for Alternatives (*Congresso Democratico das Alternativas*, CDA) in October 2012, changed the situation after support from all movements of the platform "Que Se Lixe a Troika" (Fuck the Troika), one that organised large demonstrations in 42 cities. PCP and the trade union federation (CGTP) General Confederation of the Portuguese Workers put aside previous differences and both parts of the left, the CDA, CGTP and several smaller groups, decided to appear together in anti-austerity protests starting on 2 March 2013.

Greece, Syntagma and ERT

On 25 May 2011, a large crowd of people, informed by social media, started gathering in Athens for a rally and later occupied Syntagma square, opposite the Parliament. There was no recognisable union, party or other organisation behind the call, just people calling themselves «αγανακτισμένοι» (*aganaktismeni*), the Greek equivalent of *Indignados.* Inspiration from Puerta del Sol in Madrid was obvious, and in turn, the three SE squares, Rossio, Puerta del Sol and Syntagma together with the Arab Spring, inspired worldwide "occupy" movements in New York, London, Frankfurt and elsewhere. The Syntagma occupation was in clear opposition to austerity measures imposed by the Greek elites and their Troika allies. Participation was massive, sometimes reaching 200,000 people during the evenings, and was initially peaceful. Later, clashes with the police took place and were covered by the international media. As others have also observed (see Kaika and Karliotas, 2016; Kousis, 2016), there was a clear political and territorial split between "upper" and "lower" Syntagma. In the former, the crowd was mixed, including older, conservative and right-wing citizens with Greek flags who made their presence visible by shouting nationalistic and sexist slogans against the Troika, the Greek political elites and parliamentary democracy. They stayed during the afternoon, blocking the street in front of the parliament building and left at night. In the "lower" Syntagma, the actual occupation was activated by young people, politicised and apolitical alike, some of who were experienced left activists. Slogans and banners were against the Troika and

Greek elites but also addressed at more political, anti-capitalist issues. The movement stayed alive for four months and was gradually organised in sections, such as the daily direct democracy assembly at night, an information and media centre, an international solidarity centre, first aid, food and medical supplies, security, cleaning, entertainment, organising thematic debates etc., all hosted in tents where people stayed in shifts overnight. Syntagma square mobilised ad hoc feasts, music performances, theoretical debates and theatrical acts and happenings, often used as resistance against riot police attacks (Stavridis, 2017). In mid-August 2011, on a day when fewer people were in the square, the police mounted a brutal attack, destroying the "tent city" and arresting a dozen activists. Several attempts to reoccupy the square failed after turning the whole area into a real battlefield with tear gas, Molotov cocktails and crash-flash bombs.

Two years later, in June 2013, the right-wing coalition government closed down the Greek public broadcaster ERT (in Greek), giving just six hours' notice before screens cut to black, with "no signal" (see Box 6.3). This unprecedented, undemocratic action mobilised 10,000 demonstrators inside and outside ERT's headquarters, blocking the arterial highway, Mesogeion in Agia Paraskevi, in the outskirts of Athens. It was an off-centre urban movement that lasted six months, this time occupying a large public building and its open space.

Box 6.3 "No signal", occupying public broadcaster ERT in Athens

For many, the closure of ERT became the symbol of what was happening across Greece. ERT had been blamed as a place of wastefulness and corruption by the private TV stations for years, and the government underestimated the popular revolt. Refusing to accept such an anti-social and undemocratic move, journalists, technicians, musicians and the administrative and cleaning staff, occupied the building and continued to provide content in defiance of the government. With the help of the (EBU) European Broadcasting Union, programming was provided via an online live-stream. Inside ERT, no one was paid, but no one went back home. Outside the building, in its gardens, every night there were artistic, theatrical and musical performances, with crowds of supporters coming to defend the occupation from an assault by the riot police. As a cellist said, "...we play music to entertain our supporters and to keep our morale high. Music against oppression". The undemocratic shutdown and the occupation attracted extensive coverage and solidarity from national and international media. There was even a condemnation by the Council of Europe who said, "...pulling the plug on ERT dealt a heavy blow to a fundamental pillar of democracy". Early in the morning of 7 November 2013, riot police cleared the building, arresting 24 journalists. The Syriza government reopened ERT in April 2015.

Italy

In Italy during the crisis period, the emergence of several direct social actions took place (Bosi and Zamponi, 2015). There was not the degree of momentum seen in in Puerta del Sol and Syntagma Squares, rather they were the continuation of several forms of activism and union-led demonstrations. Among these, I can mention critical consumerism and purchasing groups, time banks and occupation while self-management of factories reappeared as well as occupation of closed theatres and cinemas. Also, housing occupation and squatting drastically increased in large cities such as Rome, Milan, Turin, Florence and Napoli (Siatitsa, 2014; Bosi and Zamboni, 2015). To these, we should add two important urban-driven socio-environmental movements, against the high-speed train (NO TAV) near Turin and against illegal waste in Campania (Armiero and D' Alisa, 2012). Housing occupations are particularly widespread and reached importance in terms of scale and public attention from March 2010 onwards due to the economic crisis, as is shown on the website of the national, *Abitare nella Crisi* (www.abitarenellacrisi.org).[10] Migrant families, poor Italian families, homeless youth and activists from post-autonomist social centres – squatters whose political culture is in continuity with *Autonomia Operaia* – are the principal groups engaged in housing occupations. In this respect, the mix of housing needs in deprived social groups leads to political squatting with radical, anti-capitalist demands and fighting against evictions.

The practice of occupation is particularly widespread among artists who, between 2008–2013, occupied and self-managed several cinemas, old factories, theatres[11] and other abandoned spaces. In 2008, the workers of a Milan steel mill occupied their workspace to stop the owners from taking away the machines. The most famous case is probably that of *L'Isola dei Cassintregrati* (The Island of workers on redundancy payment) – using a similar title as an Italian reality TV show with celebrities on an exotic island. For 15 months, in 2010 and 2011, a group of chemical workers occupied an abandoned prison on the island of Asinara, off Sardinia, launching a media experiment with their individual and collective stories, which attracted intense national and international attention (Nurra and Azzu, 2011).

Finally, and moving out from the four countries, I feel obliged to highlight that along with these "local" demonstrations and occupations of squares in SE, there were hundreds others across Europe, supporting and providing solidarity to SE demonstrators and occupiers and particularly to Greece. Perhaps they were sporadic, ephemeral and attracting few people. The fact is, however, that in almost every major European city, coordinated transnational union actions by students, evicted people and comrades from left-wing parties and anarchist organisations provided immense political and moral support to protests against all these years of austerity. Of particular importance was the "Blockupy" movement that organised anti-austerity demonstrations against the ECB in Frankfurt, an emblematic action of

North-South solidarity. Organised by German activists and others across Europe every year since 2012, it faced exceptionally brutal reaction from the German riot police in 2013 and 2014, and activists responded by setting fire to several cars. "Blockupy's" solidarity finds its antithesis in German reactionary union leaders, like Berthold Huber of IG Metall, who produces divisive language blaming unions in Spain, Italy and Greece for the crisis. Unlike German unions, the Italian CGIL, in collaboration with the cultural group Associazione Arci, participated in many solidarity demonstrations and sent political delegates to Portugal and Greece. Intellectuals, academics, politicians, artists and others, from A to Z, signed dozens of solidarity petitions, often backing them with their personal presence in demonstrations, meetings and conferences, proving how important radical international solidarity is to "local" struggles. The same is true for progressive and radical institutes, such as the left European network *transform!, the EuroMemo Group* of progressive economists and the international movement Attac that organised solidarity conferences and meetings, diffusing specialised knowledge regarding alternatives to the crisis[12] among intellectuals and activists. This international solidarity was greatly needed at difficult moments, and everyone appreciated this comradeship and solidarity. As David Harvey said in a panel at the occupied ERT in Athens, addressing a huge audience – being there I quote him from memory-"...I am here for personal reasons because your struggle counts towards my struggles at home, as it counts towards other struggles in other places. We fight similar enemies in different places, at different scales and with different means, but these struggles should unite us all".

Solidarity, alternative and informal social movements

Long before the crisis, particularly in Italy and Spain, thanks to cooperative, anarchist and communist traditions, many people were already engaged in various co-operative forms of housing, production, consumption and exchange. Operating within the dominant system of capitalist relations, they developed alternative practices with the aim to go as close as they could to a use-value economy (Conill et al., 2012; Bosi and Zamponi, 2015). Before the crisis, in Italy and Spain, occupied social centres, housing and theatres, apart from securing shelter to those in need, provided self-organised education, health care and cultural activities. At the same time, as Vaiou and Kalandides (2016: 458) pointed out for the case of Greece, "...they posed broader and longer-term questions of how to live together, how to perceive and claim social needs, how to deliver goods and services, how to resist austerity and how to participate and do politics". Outside urban areas, alternative ecological agricultural production and distribution made possible alternative modes of food production and consumption. The severe housing crisis in Spain and Italy gave rise to widespread radical social movements that mobilised hundreds of activists engaged in several innovative urban actions (Siatitsa, 2014; Eizaguire et al., 2017). The crisis found these social movements and alternative ways of life somehow prepared to adapt to new harsh times and strengthen the importance of the movements. The experience gained became a prototype for the new crisis victims since the weak state and municipal welfare institutions in SE had been all but destroyed by austerity policies. Parallel to these initiatives, non-state charity organisations, mainly of the Catholic Church, and many NGOs were active in support of thousands of people in need.

The study of solidarity activities and alternative action organisations during the crisis is the theme of the extensive cross-national research with the acronym LIVEWHAT (2016).[13] The study focuses on nine EU countries, among which are Greece, Spain and Italy but not Portugal. The nine teams of the consortium studied alternative forms of coping with the crisis using the term "resilience" and taking as a unit of analysis what they call (AAOs) Alternative Action Organisations. These are formal and informal organisations operating outside corporate, state and EU agencies, including charity organisations of the Church and various NGOs (see Box 6.4).

Box 6.4 Types and groups of AAOs as in LIVEWHAT research

According to the research report, AAOs belong to seven types: informal and protest groups, social economies, NGOs, charities and church, unions/associations, municipalities/regions and other. They are engaged in a large repertoire of actions that include barter networks and swap bazaars, education and creative actions, food banks, social groceries and soup kitchens. In the field of credit, they include credit unions, ethical banks and alternative currencies. Health centres, social medicine and assistance to vulnerable groups plus advisory, consultation and psychological support form another important sub-set of AAOs. Finally, humanitarian and voluntary architecture, construction and shelter provisions, self-managed community initiatives and other actions of resilience are also included. These actions are further categorised in 10 groups of which basic/urgent needs seem to be the most important and comprises: shelter/housing, anti-eviction, soup kitchens, social grocery, health/social medicine and mental health, social support/help line, clothing, education, anti-taxation, refugee/migrant support, energy, free legal consulting and volunteers call.

Source: www.livewhat.unige.ch/.

Focusing on SE, a particular pattern exists for the highly affected countries (Greece, Spain and Italy) that is different compared to Central-North European countries. The pattern encompasses, first, the highest frequencies of informal and protest groups at 44 per cent–47 per cent (among the seven types mentioned in Box 6.4), followed by NGOs and social economy.

Second, almost half of all AAOs in SE were established in the period from 2008–2015: 56.2 per cent in Greece, 50.4 per cent in Spain and 44.8 per cent in Italy. And third, actions related to basic/urgent needs, social economy, alternative consumption and culture are the priorities of AAOs in SE.

Unsurprisingly, Greece holds the highest frequency in basic/urgent needs compared to other countries and specifically in the provision of clothing/items, free healthcare/medicine, soup kitchens, refugee/migrant support and actions against taxation. As the research underlines:

> ...These reflect the intensity of the impacts of the financial crisis and the related dramatic decline in the standards of living for a considerable part of the Greek population, as well as impacts of the recent refugee crisis in Greece.
>
> (p. 59)

Likewise, Spain is high in basic needs and leads the anti-eviction actions but shows also high frequencies in non-basic/urgent needs, cultural activities, alternative consumption and social economy. Italy does not have high frequencies in basic/urgent needs but does in alternative consumption activities, self-organised spaces and social economy. Finally, in terms of the beneficiaries of AAOs, the uninsured, unemployed/precariously employed and refugees/migrants constitute the majority in Greece. In Italy, it is the general public, citizen-consumers/SMEs and local communities while in Spain it is poor or marginalised people/communities and the general public.

Although charity organisations are specialised and effective in helping those in need, as documented by LIVEWHAT research, I want to focus instead on informal, self-organised, bottom-up solidarity initiatives, following Galeano's suggestion at the beginning of the chapter. Informal networks of solidarity, what Rakopoulos (2015) calls "hidden welfare" in the case of Greece, find their roots in the wider tradition of informality in SE, discussed in Chapter 2. I remind readers that informality is a structural characteristic of SE societies with important spatial and sectoral differences, and it is embedded in the formal institutional system. It takes new, urgent forms due to crisis hardship effects, bringing back, as Papataxiarchis (2016c: 17) suggests, "...sociality on 'horizontal, anti-hierarchical' terms". These solidarity initiatives have the potential to construct a counter-austerity narrative and to contribute towards transformative relations that are a central practice of the political left. Or, to recall Gramsci, they could build among the subaltern SE people a counter "philosophy of praxis", different from the top-down welfare institutions of the state, the Church, charity organisations and NGOs. The solidarity networks that I will speak about should be understood as highly politicised and initiated by progressive collectives, some of which are of leftist and anarchist origin. In this respect, their aims are to promote equality, dignity and socio-spatial justice. Along these lines,

Featherstone (2012) reminds us that solidarities are constructed through uneven power relations and geographies and remain fragile, particularly those associated with the left. In this sense, solidarity could be part of the process of politicisation and alternative political emancipation.

The coincidental timing of the economic crisis with the housing crisis, in Spain, leading to widespread evictions, and of dispossession of public housing and rent increases in Italy enforced by insurance and pension funds, made shelter a basic and urgent need for thousands of people in both countries. It was the most crude and immediate result of the real estate bubble in combination with the aggressive banking policies towards loans, the explosive moment of the contradiction between the use value and exchange value analysed by Marx. At the beginning, with falling housing prices, people lost the exchange value of their houses. In the end, they also lost the use value of their own dwelling houses (see also Harvey, 2014). Unlike in Greece, in Italy, eviction victims and marginal people without housing access participated in the massive occupation of empty buildings so that by 2014 more than 100,000 people were living in occupied spaces. Parallel to occupations, actions supporting evicted persons took place. In Spain, the Platform of Mortgage Victims movement (PAH, Plataforma de Afectados por la Hipoteca) emerged in 2009 in Barcelona and later extended to all major Spanish cities. It was a bottom-up, self-organised movement initially of owners who had lost their houses and joined later by other activists (Observatorio Metroplitano, 2013; Eizaguire et al., 2017). The movement succeeded politically, not only in assisting dispossessed people, but also in uncovering capital's greed and injustice in the neoliberal urbanisation process. The PAH solidarity movement gained substantial coverage in national and international media and was engaged in collective actions to stop evictions, protesting in front of banks and regional/municipal governments and fighting with legal means to change housing and eviction laws (see Siatitsa, 2014; www.afectadospor-lahipoteca.com).

The housing crisis in Italy accelerated after 2010, and occupations multiplied, although in some old "left-wing cities", such as Bologna, new municipal authorities of the "Third Way" evicted several occupations in the period up to 2015. Long-standing movements such as Coordinamento Cittatino di Lotta per la Casa and Blocchi Precari Metropolitani, plus occupied social centres, were engaged in housing squats (Grazioli, 2017). The Platform Abitare nella Crisi, mentioned before, coordinated in Rome, in December 2012, a simultaneous occupation of ten buildings with 3,000 participants (Siatitsa, 2014). Those who find shelter in these solidarity occupations are mainly migrants, but unemployed and homeless Italians are also included. Occupations after 2014 faced a strict law that criminalised everyone who "illegally" occupied a building citing the lack of a legal address necessary to find a job, to get health care and to put their children into school.

Since 2010, the crisis has generated hundreds of new social solidarity initiatives across Greece, which, as noted by Kousis et al. (2016), Rakopoulos

(2015), Cabot (2016) and Vaiou and Kalandides (2016, 2017), are mainly engaged in the provision of urgent/basic needs. Solidarity actions known from other countries and other emergency services such as social kitchens, social groceries, local and national exchange networks, local alternative currencies and time banks have become widespread in urban areas[14] (see also Sotiropoulou, 2012; Vathakou, 2015). As the above authors underline, these solidarity actions vary considerably depending on the social group and the community they target, the needs they aim to cover, the relationships among volunteers and the relationship with institutions. The majority are usually hostile to the state and NGOS while some of them, such as the Hellinikon social clinic, accept help from progressive municipalities. In the following section I look more closely at three Greek cases, first distribution of "food without middlemen", second "self-organised social clinics" and third support to "migrants and refugees" to show how austerity politics at home and the effects of imperialist wars in the Middle East and Africa could be interconnected and in turn at least partially contested via solidarity actions. There is by now extensive research on these issues and what follows are short summaries of other's research plus my own sporadic and non-systematic fieldwork, used to illustrate my general argument.[15]

Food "without middlemen"

Buying food "without middlemen" is a movement applying alternative economic action that came out of the urgent need to supply food to Greek society at low prices. Incidents of hunger, people searching for food in garbage bins and in leftovers from open farmers' markets, increasing demand for soup kitchens and children fainting in classrooms due to hunger became everyday episodes, re-enacting sad memories of the Nazi occupation during the Second World War. These extreme conditions for a Eurozone country found the state and municipalities unprepared. Apart from charity organisations engaged in soup kitchens, the movement distributing food "without middlemen" acquired an important position in the food distribution chain. It started as the distribution of potatoes at a low price directly from farmers to consumers in March 2012 in Katerini, a town in Northern Greece – hence the movement is also known as the potato movement – and soon expanded to include a large variety of food products[16] (*Avgi* newspaper, 11 March 2012, in Greek; Calvàrio et al., 2016).

According to Skordili (2013), the recent food poverty, apart from the crisis and the corresponding unemployment and reductions in income, relates also to a major restructuring of the grocery-retailing sector. The increasing power of corporate retailers and the destruction of small-scale neighbourhood groceries resulted in 2011, as Skordili (2013: 134) argues (using data from the Greek statistical service), "...in increasing the consumer price index for food and non-alcoholic beverages by 4.3% while the average index for all products increased by only 2.4%". Thus, food prices went up while demand was decreasing, and food prices in the large corporate retailers dominating the market in Greece were considerably higher than in richer countries like Germany and France (*Eleftherotypia*, 19 October 2011, in Greek).

The movement contested the role of wholesalers as well as corporate supermarkets. In doing so, it attracted support both from farmers suffering from imposed low prices in the fields and consumers suffering from high prices in shops. Rakopoulos (2015) argues that Greek groups engaged in the anti-middlemen movement do not aim to challenge the money exchange system. He underlines the importance of a solidarity economy and its affinity to informality and adds:

> ...Reciprocity and market exchange co-exist in the anti-middleman movement of Greece and there is a strong affinity between the solidarity economy and informality. More profit for producers, less cost for consumers, sociality that binds activists together in the formation of markets – all of these work together to create an integral realm.
>
> (p. 98)

Food distribution actions in various places initially attracted many people. They were organised by solidarity groups and enabled farmers to sell their products directly to consumers at pre-agreed prices, usually 20 per cent–50 per cent lower than the standard shop price (Calvário et al., 2016). The idea rapidly spread across the country, especially in Athens and Thessaloniki, where the effects of austerity were more severe. The products diversified to include flour, vegetables, olive oil, olives, honey, feta cheese and others. In 2015, according to Solidarity for All, a network connecting and supporting solidarity groups, there were at least 40 "without middlemen" solidarity groups across Greece, 23 of which operated in Athens. They succeeded in helping 20 per cent–25 per cent of the Greek population, distributing several thousand tons of food (www.solidarity4all.gr). During the first years of crisis, on many occasions, farmers distributed excess production of fruits and vegetables free, attracting huge crowds. Photos depicting dozens of people with raised hands waiting for a few tomatoes were widely published and circulated in social media.[17] According to Calvário et al. (2016: 11):

> ...solidarity groups self-organize through open assemblies and consensus procedures. They informally coordinate at the regional level, with five national events organized between 2012 and 2015, three in Katerini and two in Athens. More than 5000 tons of food was distributed between 2012 and 2014.

A substantial number of other solidarity groups periodically organise food distribution "without middlemen". Among them are: a social grocery business in Xylokastro, Peloponnese, doing also cultural and educational activities; a time bank in Athens that also organises free foreign-language training and legal support; and a women's collective in Thessaloniki that runs a small restaurant but operates also as a free educational centre and hub for ethical trade. Solidarity organisations choose to have economic ties, such as the buying of consumables and other supplies with initiatives operating with similar practices. The issue of food poverty is directly linked with questions about social justice in the uneven development that hit Greece (Skordili, 2013). Buying and distributing food "without middlemen" shows how effective the direct connection between producers and consumers can be. It shows also how social intervention in the sphere of distribution and consumption of basic food can change relationships, not only that between producer and consumer, but also the "...value and meaning, rather than solely the type of buying of the product itself" (Marsden et al., 2000).

After 2015, many solidarity groups continued their efforts, but demand has been gradually decreasing. The Syriza-led coalition government introduced two important actions to ameliorate food poverty. First, the prepaid Solidarity Card that can be used by people in extreme poverty, around 250,000 across Greece, to buy food up to 220 euros per month. Second, the introduction, for the first time after the Second World War, of school meals, initially to a small number of schools and from 2016 onwards, after a donation from the Stavros Niarchos Foundation, to more than 300, located in the most deprived regions and cities of the country. Although these two actions are very important, they failed to disrupt the established food chain and to promote an alternative, progressive and ecologically friendly mode of connecting farmers and consumers. They failed to create spaces for local initiatives to find ways of using procurement policies for food purchases, not only for schools, but in a variety of prosaic settings such as hospitals, the army and refugee camps. In other words, they failed to improve the quality of food and thereby deliver a variety of health and well-being benefits, as well as stimulating local agricultural economies as happened in Italy, France and the UK (Hadjimichalis and Hudson, 2007).

Self-organised social clinics

The most unique solidarity action in Greece, and perhaps in SE, is the *social solidarity clinics and pharmacies*, a direct outcome of the decimation of the Greek NHS and the introduction of the 2012 law prohibiting access to public hospitals to uninsured Greeks and migrants. The first social solidarity clinic in Rethymnon, Crete, 2008, was established with the aim of providing health services to migrants. It was followed by the second in Thessaloniki in 2009. After the Syntagma Square movement, some of the doctors from the square's first-aid unit organised the third social clinic and pharmacy in Hellikon, Athens. Today (2016), there are 53 solidarity clinics functioning in Greece, 22 in Attica, four in Thessaloniki, two each in Rethymnon, Heraklion, Larissa, Patra and Ioannina, and others in different cities, including on large islands. As the (SSCP) Social Solidarity Clinics and Pharmacies Charter (2013) states:

> ...Social solidarity clinics and pharmacies are self-managed, autonomous and independent collectives of people who voluntarily and completely free of charge, provide medical and pharmaceutical care to the people deprived of social/medical security coverage (uninsured), those in need and/or unemployed, Greeks and migrants, without discrimination, regardless of religion, nationality, sexual orientation, gender and age.

Throughout the years of austerity, over 3,000 volunteer doctors, nurses and technical and administrative personnel provided primary healthcare in Attica only. As with the rest of the solidarity movement, SSCPs provide cost-free services, and everyone works on a strictly voluntary basis. All equipment – often expensive and very modern – medicines and consumables come from donations by ordinary people and doctors, while many equipment donations come from other solidarity groups across the EU. The campaign of medicine collection and sharing among SSCPs has become so successful that in quite a few cases solidarity pharmacies have informally provided medicines to public hospitals via doctors who work in both social clinics and hospitals. While the movement avoids charging money, there is a need to cover utility bills and the medical consumables for everyday use that are not donated. International financial solidarity remains invaluable and comes from various European, US and Canadian solidarity groups who "adopt" an SSCP, taking care of its bills. In addition to solidarity actions in Greece, the network of SSCPs organised two missions to the Gaza strip (2013 and 2014) and one to Kobane (2015).

The crisis changed dramatically the composition of beneficiaries. As a doctor in one of the first social solidarity clinics in Thessaloniki said:,

> ...initially we helped mainly migrants and very poor people from the town and rural areas. From 2012 on, the majority of our patients were middle class unemployed Greeks (...). This changed after 2015 when Syriza abolished the inhuman law (...). Now the majority are migrants and refugees again.
>
> (interview by the author, March 2015)

The Thessaloniki social clinic participates in a variety of solidarity actions apart from health care. Among these I could mention the organisation of children's playgrounds, the support to the movements against water privatisation and against the gold mines in Halkidiki and fighting xenophobia in schools in Northern Greece.

In her research in two Athenian social clinics, Cabot (2016) argues that these solidarity actions should not be framed as radical breakaway, but rather as operating within the injustice framework that arises from privatisations and austerity. She highlights that efforts by volunteers reconfigured the "...relationship between individual, social and political-economic bodies-in-crisis" (p. 155). And she concludes:

> ...The social clinics present alternative approaches to social and individual bodies-in-crisis by reconfiguring the distribution of labour, care and medicines, addressing not just individual somatic needs but also social and economic relationships. The social clinics thus produce new paradigms of citizenship which do not fit easily within more entrenched structures of belonging in Greece, such as the family or the nation-state; and they strive explicitly to undermine the marginalizing, violent work of an increasingly austerity-driven Eurozone
>
> (p. 156)

Many social clinics try to operate differently from other similar NGOs, such as "Doctors Without Borders" and municipal services. They try to engage the beneficiaries in the everyday operation of the clinic, to bypass the dominant pathetic recipient status of patients in state-run hospitals. Maria, from the Athens self-managed solidarity clinic, describes the difficulties in applying these principles:

> ...we started in a small apartment in Exarchia (...). With donations from French and German comrades we moved to a larger place close to Omonia (...). All here work as volunteers because we believe strongly in solidarity. Charity makes people apathetic and patients think that we are a state or municipal service. They don't believe that we are self-organised and that everyone works for free (...). It is difficult to convince these ex-middle class and now marginalized people to provide help; it challenges their 'ex-' status. With poor people and migrants it is easier
>
> (interview by author, June 2015)

Nevertheless, volunteers continue their efforts, and as a first step, all patients get informed about the organisation of the clinic as a self-organised solidarity initiative, run by unpaid volunteers and with limited facilities. Second, all patients are asked to return unused medicines and/or collect more from friends to donate to the pharmacy. Those patients and their relatives who are able to help are encouraged to participate in simple everyday tasks, such as cleaning, acting as translators, helping the elderly to return home, playing with children, etc. Finally, some of the former beneficiaries became volunteers themselves in administrative tasks after a short initial training.

Box 6.5 The Athens social solidarity clinic

At the Athens Social Solidarity Clinic, 26,743 cases were taken care of from February 2013 until December 2016: 13,424 patients were examined on the clinic premises and in private practice by volunteers; in 10,649 cases, patients received medicine; and in more than 2,670 cases patients were helped to get an appointment at the public health structures. On average, 500 patients contact the clinic monthly (The clinic annual report, 2017). Since June 2016, and especially since September 2016, the number of patients who come in order to get examined or get their medicines is decreasing significantly, since all those who have a social security number, regardless of their nationality status, now have access to the public health system. Twenty-six doctors, four pharmacists, four technicians and 65 volunteers work in shifts at the social clinic, a rented office space close to Omonoia Square. In addition, there are 16 collaborating specialists who provide health care and specific tests in their own premises, accepting a number of the clinic's patients per week without charge. Among doctors, men and women are more or less equally represented, but among the volunteers, all are women apart from three men. In the group of doctors, middle age predominates, while among the volunteers, the great majority are pensioners. Finally, in terms of education, the majority hold a university degree. A network of German, French and Swiss doctors and individuals supports the social clinic by paying its rent and utilities bills, supplying equipment and medicines and working periodically as volunteers.

"Refugees-migrants welcome", a crisis-within-a-crisis

During summer 2015, a wave of thousands of refugees-migrants from Syria, Afghanistan, Iran, Pakistan and Somalia crossed the short sea pass from Turkey to the neighbouring Greek islands.[18] In one year, more than 1,200,000 men, women and children tried to escape to Europe via Greek shores, leaving behind many who lost their lives in the sea. While Kos, Chios and other Greek islands in the Aegean have also received refugees and migrants crossing over via Turkey, Lesvos has received the highest number: more than 93,000 in three months in 2015, which is more than seven times the 12,187 arrivals in all of 2014. Lesvos has just 30,000 inhabitants, and this huge influx dramatically changed their daily life. According to (UNHCR) United Nations High Comissioner for Refugees, 500,018 people arrived in Lesvos from January to 31 December 2015, 46 per cent men, 34 per cent children and 20 per cent women, to continue their journey to Europe.

Lesvos is a major migration entry point to Europe, at least since the 2000s. Together with the thousands of refugee-migrants, it has attracted several bottom-up solidarity groups, which are helping people coming from the "East". In 2009, radical-left solidarity groups organised an international "no-border camp" and several activist actions across the island (Lafazani, 2012; 2014). An umbrella of local organisations, with the support of the Church, "Doctors Without Borders" and students/staff from the Aegean University, decided to reuse a children's camp as a refugee-migrant centre, after several racist attacks on migrants. They named it "to Χωριό του Ολοι Μαζί" (The All Together Village).[19]

Although the island had experience in receiving and helping hundreds of refugee-migrants, the high numbers arriving in the summer of 2015 found it unprepared. The same is true for the local municipality and the Greek state that were unable to deal effectively with such a big problem on their doorstep, and soon the situation became a "crisis-within-a-crisis". During the first days, rescuing people in the sea was undertaken by local fishermen and later by NGOs, professional rescue boats and the Greek Coast Guard. On the land, local people welcomed the new arrivals, providing water, food, dry clothes and information on how to go to the port of Mytilini to take the boat to Piraeus. Local people, while they helped to rescue migrants, as Papataxiarchis (2016a) accurately pointed out, said they were "doing nothing". Their statement meant that they were doing what they always did, helping people in need, "doing what has to be done". Papataxiarchis (2016a) continues, describing his experience in Skala Sykamias, a little port on Northern Lesvos, an "informal gate" to Europe due to its proximity to Turkey. He analyses the "doing nothing" response of local people to a journalist through the act of "three grannies", captured by a photographer, feeding a refugee baby. He writes: "...Feeding a refugee baby is more than 'business as usual', but a historical opportunity for an elderly woman to be energetic agent in the only – i.e. culturally specific-way possible: that of looking after children" (p. 9).

There were, however, other "solidarity" groups in Lesvos. First, there were the international NGOs that came after the refugee crisis became known. These are professional teams, well organised and equipped with a central hierarchy and specific goals. Second, was the UNHCR, the UN agency for refugees that coordinated NGOs while providing equipment and personnel. Third, professional but unpaid volunteers formed rescue teams in the sea with speedboats. Fourth, and isolated from the previous categories, were highly politicised groups from the radical left and anarchists groups, Greeks and foreign alike, who occupied public open spaces, building their own sectarian solidarity communities providing food and shelter. And fifth were temporary volunteers informed by social media, who came for a few days to help. Apart from solidarity organisations, there was the Greek coast guard that rescued more than 1,400 migrants in nearly 60 search and rescue operations near several Greek islands in the Eastern Aegean Sea over a three day period in August 2015 (*Efimerida ton Syntakton*, 16 August 2015, in Greek; Rozakou, 2015). I noticed, also in Lesvos, only a few sporadic xenophobic reactions to refugee-migrants and to solidarity groups, from a minority of locals, while in Chios and Kos, the participation of Golden Dawn racist activists escalated opposition to clashes with volunteers and the police (*Avgi* newspaper 20 August 2015, in Greek).

Lesvos as a transit island survived, albeit with lots of difficulties and tensions during the refugee crisis of the summer of 2015, because the Balkan road from the Greek border to Munich was open. The vast majority of the refugee-migrants do not wish to stay in Greece, as there are no prospects, and want access to richer EU countries by either walking across the Balkans from Northern Greece, or sneaking onto Italy-bound ferries from the West. After the construction of fences by various European countries, and particularly the fence at Idomeni, the Greek border pass to FYROM(Former Yugoslav Republic of Macedonia), the Balkan route closed, and thousands of refugees and migrants were trapped in Greece, many on Lesvos and other islands. As a report from Amnesty International (2015) pointed out[20]:

> ...This is not just a Greek tragedy, but a Europe-wide crisis. It is unfolding before the eyes of short-sighted European leaders who prioritize securing borders over helping survivors of conflict. The world is seeing the worst refugee crisis since the Second World War. What Europe's borders need is not fences but safe entry points for refugees, and facilities to receive them with dignity.

From the beginning, the Syriza-led coalition government, unlike others in the EU, with the exception of Germany, had a very pro-migrant attitude and much more open policy towards the refugee-migrants. Referring to local people at Lesvos, several high-ranking officials, including PM Alexis Tsipras and the President of the Republic Prokopis Pavlopoulos made statements such as "this is the image of Europe that we want" and "they have saved Europe's humanitarian face and values" (*Avgi* newspaper, 18 August 2015, in Greek), what Papataxiarchis (2016b: 4) ironically calls "a new patriotism of "solidarity". The situation, however, radically changed after the EU-Turkey agreement on 20 March 2016. On the one hand, the number of incoming refugee-migrants dropped sharply, an indication of the geopolitical games being played by Turkey. On the other hand, the agreement introduced a new role for Greece as a "buffer zone" between Turkey and the EU, where these people would have to stay in geographically diffused camps, so-called "Hot-Spots". But stay for how long and waiting for what exactly? These questions remained unanswered, and from a welcoming attitude to refugee-migrants, the Syriza-led coalition moved to enclosures of people in fenced and guarded "Hot-Spots", and the whole situation shifted to a gradual militarisation with the presence of Frontex and NATO, while accepting Turkey as a "secured" country for deportations.

The end result was 60,000 trapped refugee-migrants in Greece by the end of 2016. Under these conditions, housing became an emergency issue, especially as the institutional response was opening refugee camps and detention centres in remote areas, where in some of them, people lived in harsh conditions.

To pressurise the government to accommodate a few families with children and individuals in Athens, the Network of Social Support to Migrants and Refugees, "to Diktio" mentioned before, together with other radical groups, organised, in September 2016, the occupation and squat of the City Plaza 8-floor Hotel, which had been abandoned since 2010. Located in the Athens centre, the occupied hotel soon became a symbol of an alternative, grassroots, informal and self-organised multicultural co-habitation (Deltio Thielis, www.diktio.org/node/1261). "To Diktio" has long activist organisational experience in supporting migrants in Athens, Lesvos and other places, including taking care of the 300-migrant hunger strike in 2011 along with other solidarity volunteers (see Mantanika and Kouki 2011; Hadjimichalis, 2013)

According to Olga Lafazani, an activist at the occupied hotel, "...the City Plaza squat is probably – for everyone who participates in it-a unique experience, very different from other political and activist actions" (discussion with the activist, February 2017). It brings together and spatialises, in one particular building, different struggles and themes, from local to global, in clear opposition to institutional policies. The translator A.R.T. in City Plaza, a migrant himself, describes how they did it and his feelings(see: Deltio Thielis www.diktio.org/node/1258):

> ...the organisation part (of the occupation) was not easy, but everyone seemed to enjoy it. After six months of self-organized and collective work, we pulled down the fences and borders between nationalities. We now feel stronger having shown to government and NGOs, how a real house without frontiers can be built.
>
> (my translation)

After an initial period of hostility, the neighbourhood seemed to have accepted the reality of the occupation, thanks also to the support by progressive media and despite the fact that there had been racial attacks by Golden Dawn in neighbouring Victoria Square in the past. Volunteers and people living in this peculiar "village" regularly clean the area around the hotel, while the 400 residents create demand that benefits the local shops. The nearby school organised a common celebration with Plaza's children and after the Syriza/(UNICEF) United Nations Children's Fund initiative for the schooling of migrant children, several children from Plaza Hotel go to public schools, in collaboration with a radical teachers' union. The volunteers, however, faced several difficulties in building this alternative co-habitation. As they describe, perhaps the greatest challenge was to convince people without previous experience that Plaza is an unfunded and self-organised project, and all have to contribute with their work. Olga Lafazani describes a particular event (see Deltio Thielis, www.diktio.org/node/1258):

> ...For most people it is profound to be in solidarity with family or close friends but not with people that you have never met before. In this sense

> the "understanding" does not come from words but from actions... For example the morning that the water company tried to cut the water in City Plaza and we all went down together to prevent it, this action was much more "convincing" for the residents, who saw that City Plaza is not a given, institutional space, but a space for which we have to fight all together to support and maintain. (my translation)

The City Plaza hotel is a hopeful and so far successful example of an informal "institution", self-organised and anti-hierarchical, providing a solution to the urgent need for housing, food and security for about 400 refugee-migrants. The question of up-scaling, i.e. whether the City Plaza could be used as a model for the accommodation of the 60,000 trapped people in Greece, remains unanswered despite its economic efficiency and above all its social and cultural sensitivity.[21]

New political subjects, emancipation and the spatialisation of politics

The need to go beyond an idealised and romanticised reading of the previously discussed protest, occupation and informal solidarity movements is obvious. Some analysts see these movements only as the result of pure spontaneity, digital organisation and transnational activism, while from their practices, they highlight only the occupation of the squares, direct democracy and the "multitude" of Hart and Negri (Vradis and Dalakoglou, 2011; Fernádez-Savater et al., 2017). Equally, I move beyond the reading by ruling elites, the mainstream media and the academic establishment, who see these movements as apolitical bubbles without social significance, since no leaders or parties are involved (a position also accepted by several leftists). I also distance myself from some hyper-politicised romantic views about the revolutionary potential of these movements and their sectarian practices. Many times reality has proved less revolutionary and more complicated than anticipated. And finally, I try to avoid a simplistic link between the multiplicity of individuals, groups and initiatives in these movements and some left-wing parties (such as Syriza and Antarsya in Greece, Bloco in Portugal, Rifondazione in Italy, the United Left (Izquierda Unida) and Podemos in Spain and various green parties) despite the active participation of leftist activists from these parties. Although there is no path-dependent trajectory connecting anti-austerity movements with the past, many activist practices since 2010 repeated old-known repertoires; a good number of persons in their 60s and 70s were there; the key role of socio-spatial contradictions emerged again; and, as in the 1970s and 1980s, many prominent activists have been elected in national/regional governments and municipal elections.

Protest and occupation in Rossio, Puerta del Sol, Plaza Cataluña, Seville, Syntagma, Thessaloniki and in other SE cities, unlike the multitude of Hard and Negri, were not abstract social categories, but a material and bodily coming-together from all parts of the city in central and symbolic places. There is a lot of debate, however, about how "inclusive" and "representative" the squares' occupation was of those in need. Younger middle class activists gave the tone, but the most deprived parts of the population were absent – feminist and ecological issues were marginalised and union participation was rejected. The same is true for demands by other social groups such as small shopkeepers. In this respect, claims by different researchers about the squares as "commons" seem exaggerated and not always corresponding to findings "on the ground".

Indignados, a Spanish expression, and its Greek version *aganaktismeni,* come with deeply problematic and contradictory connotations, insofar as they simply mean that people affected by the crisis show nothing but indignation, outrage and anger (Kaejane, 2011). At the same time, it seems to encapsulate a minimal but quite powerful basis for the co-existence of thousands of citizens not directly associated with some form of collective representation through unions, activist groups and political parties. A major difference between the Spanish and Greek cases is that in Greece there was active involvement (from the first day to the last) of radical left activists as individuals (from Syriza, Antarsya and other non-parliamentary left and anarchist groups), while in Puerta del Sol, the majority were angry citizens of middle class origin without a direct interest in leftist alternatives, despite the presence of activists from *Izquierda Unida*. The role of experienced leftist activists within the *aganaktismenoi* in Syntagma was contradictory. On the one hand, they were crucial in providing organisational support and shifting slogans and demands to more politicised issues. On the other, they made many people suspicious and in a few cases destroyed the climate of "being together against the Troika".

Nevertheless, what took place during those days in all the squares was the formation of a new political subject distinct from traditional forms of political representation. It was not from the unions, not from student organisations, nor from any other recognised organisations or political parties. It was a new political "morphoma", a coming together of very young persons with middle-class unemployed people, with civil servants, with precariously working people and with sections of the non-unionised working class. Many of these people were participating in demonstrations for the first time.

Some of the SE initiatives are "accommodative", others "transformative", following Nancy Fraser (1995), and no one knows how long they will last. But after all, what those cases demonstrate is a struggle to introduce alternative political hegemony over everyday life for thousands of people in poverty and despair. And other cases from beyond SE have demonstrated how the arena for alternative struggles over the production and distribution of resources among all people (including migrants and refugees), firms and institutions can be opened up (Hadjimichalis and Hudson, 2007). Similar experiences are well known worldwide from times of war and economic, social and physical disaster, such as in Palestine, in the Yugoslav war, in

Argentina, in Cuba, in Chile and now in SE countries. What makes the case of SE countries different is their position among the wealthy nations and being members of the EU and the Eurozone.

I want to argue, however, that they can be seen as emancipative if we include *the socio-spatial context and the political conjuncture*. Thirty years ago in Europe, such an approach could be easily dismissed as tangential, even obfuscatory, in the context of devising more radical policy approaches. In the present crisis, and under the ultra neoliberal attack, when whole societies and countries are turned to "territories of exception", we have to fight for what we not so long ago regarded as the basic rights of citizens in social democratic societies. And finally, these cases are occurring in a highly polarised political environment in which all initiatives are interconnected with the wider demand for radical political change, while xenophobia and neo-Nazism are on the rise. On the terrain of everyday life, solidarity movements developed to contest and politicise austerity by doing something lasting longer than a three-hour demonstration. Working "at the core of society with the subaltern", as Gramsci (1919) suggested, activists generate or regenerate self-organised and bottom-up movements that in many cases succeed in challenging the hegemonic fatalism that nothing can be done. Taken together, social movements, protest events and informal grassroots solidarity initiatives all challenge head-on the one-dimensional emphasis that runs through neoliberal policy concepts and policies. And if in the longer term these struggles could contest the imperatives of capital along with issues of social reproduction and oppressive divisions around gender, race and sexuality, they could be seen as emancipative.

The successful campaign and final victories in the 2015 Spanish municipal elections in Barcelona, Madrid and other cities among which Zaragoza, Cadiz, La Coruña and others, were positive political outcomes of previously discussed local movements. Their policies and everyday politics after two years in office signal the will towards emancipatory democratic politics. The success of the coalition "Barcelona en Comú" that won the municipal elections in Barcelona, allowed Ada Colau, the activist and informal leader of the Platform Against Eviction (PAH), to became the first woman Mayor of the city. Another case is the victory of the "Ahora Madrid" coalition in Madrid, and its leader, Manuela Carmena, who became Mayor; she is a feminist and former communist judge and a leading figure against Franco's dictatorship. The triumph of both is, undoubtedly, the result of citizens' votes being affected by the housing bubble crisis and the collapse of the fantasy that living an indebted life is without problems. It marked the biggest breakthrough yet for Spain's new left and the direct involvement at the municipal scale of former *Indignados* and other movement activists. It is premature to review whether their up-scaling attempts have been successful. Their priorities and procedures, nevertheless, are indicative. For example, in both cities, half of the municipal budget is directed to public social affordable housing, with parallel intensive negotiations with banks to stop evictions. Furthermore, both mayors reject big, speculative real estate projects, and

Madrid withdrew its candidacy to host the Olympic games. In Barcelona, the Mayor and her cabinet struggle to have every month a general assembly with movements' representatives, while on-line discussions and evaluations of particular neighbourhood projects are under way. Of course, the contradiction between the mode of anti-hierarchical collective activist work and the established bureaucratic hierarchy is there, but these two cities show, so far, that a different emancipatory and transparent governance of a crisis situation is possible.

A crucial parameter of emancipation has been the *spatialisation of politics* via the struggles mounted by the different movements. Real estate speculation and the housing crisis, particularly in Spain and Italy, combined with unemployment and poverty, have provided the material urban basis for outrage and indignation. Occupation of housing, streets and squares was a key aspect of the post-2011 movements' cycle. In public spaces, people can symbolically claim not only the right to the city, but also the right to dignity, protesting against the several socio-spatial injustices imposed by austerity policies. Regarding the 15M, Seville-Buitrago (2015) argues "...The spatialities deployed by the Indignados are complex and variegated, multiplicative, trans-scalar, and inducible to a uniform logic as a direct result of the plural, spontaneous nature of demonstrations". I think his comments are applicable to other SE squares' occupations, while bearing in mind important differences in content, participation and duration. Demonstrators turned each square into a social, political and non-violent space of solidarity, not just a ceremonial public space; a coming together with a common political desire to demand, first, their right to occupy the symbolic space of the city's most central square and to express their opposition to austerity measures; second, to demand their right as citizens to *centrality* and *self-management*, recalling Henri Lefebvre, both in urban terms (the most central city square) and in political terms, demanding dignity, citizenship rights and radical changes to the oppressive political system. The "enemy", in the case of Spanish activists, was the corrupt national political system, while in Portugal, and more so in Greece, the Troika and German politicians were also included. Despite deep unevenness, what seemed to connect the movements in the squares is that this open spatialisation both resulted from and made visible the growing de-legitimisation of the political elites and their austerity policies in each country. De-legitimisation in the squares didn't come from the usual suspects – the unions, the organised left and the anarchists – although some were there. It came from the majority of participants. These were young precarious people and the deprived and angry middle classes, who became the new, and unexpected to the elites, radical political subject (Gonick, 2016).

As Massey (2012) argues, space provides the possibilities for co-existence and multiplicity, being the material and mental support for simultaneity dependent on contiguity, unlike time, which is the dimension of succession. In the streets, squares and solidarity premises, SE activists reinvent co-existence and simultaneity, and at the same time, they inhabit the digital spaces of communication and information diffusion. Digital communication introduced new spatialisation: simultaneity without contiguity. But let's not exaggerate: technologies do not create movements. As Castells (2016) insists, movements became movements by occupying space, by being "there", visible and rebellious or by resisting in informal solidarity movements. The de-territorialised and territorialised spatialisations work well together only if the first succeeds in mobilising many people for the second and if they can strengthen each other with the speed and spread of digital communication. In this respect, the rapid spread of information through social media to enable gatherings and start demonstrations has been a useful modern tool. The clear danger is to see digitalisation as an end in itself, bypassing the questions of what action to take, how to struggle and what is to be the next step after gathering spontaneously together.

The three Greek examples, food "without middlemen", social solidarity clinics and helping the refugee-migrants, are more than a response to society's urgent needs. Through "food", "health care" and "welcoming refugee-migrants", activists seek to politicise three key issues of social reproduction, as Fraser (2006, 2013) suggested, and to move the struggle towards terrains that usually remain marginal among many leftists. The spatialisation of these actions played a key role. In "food", relational space appears as the reconstruction of the urban-rural relations and in the rebirth of the direct producer-consumer relationship. It is an encounter in an ephemeral "market place", usually an occupied square, different each time, resembling the characteristics of a festive occasion ("panigiri" in Greek). In "health care", the spatiality of social clinics combines stable premises with a changing and multi-scalar relational space that covers urban areas and extends itself to transnational volunteer networks. In the Athens centre social clinic, the majority of patients come from the city centre's deprived districts.

The moving bodies of refugee-migrants reflect and produce a contested, contradictory and multi-scalar spatiality that contains all possible applications of spatial practice, relational space and spaces of representation. By crossing political, physical and cultural borders, they have made transparent and contested spaces in "borderlands", while also generating kilometres of fence construction along EU borders to make a material reality of "fortress Europe". The newcomers from the East have transformed local native societies, generating solidarity movements where all kind of encounters between refugee-migrants and volunteers take place, with "here" and "there", with the "present" and the "past". At the City Plaza Hotel, the spatialisation revolves around its function as a multilingual/multicultural "transit village" where its inhabitants wait to continue their journey to a European country. They build seemingly ephemeral relationships, which may last forever, and they construct a moving centrality in the centre of an old city.

* * *

Summing up, these alternative ways to contest austerity's hegemony, together with all the other forms of struggle, protest and occupation discussed in this chapter, have been the most fruitful and promising outcomes from the harsh times in SE. The successful municipal examples in Spain opened up a fruitful emancipatory path. Capitalists keep saying: "crises create opportunities", having in mind their own speculative profits. For the rest of us, who know that these movements appear and disappear in time and space; the question is not why they disappear – this is inevitable – but how. What kinds of seeds have they planted in people's collective memories, and what kinds of political opportunities have they opened up? Social movements have planted their seeds in Southern soils. The question now is who and how will we do the harvesting.

Notes

1 Slogan of the demonstrators in the occupied Syntagma Square, June 2011, adopted later by the Greek network "solidarity4all".

2 There is by now an extensive literature concerning social movements and insurrections during 2011 in SE. Among the many see: Vradis and Dalakoglou (2011), Davis (2011), Harvey (2012), Douzinas (2013), Kaika (2012), Hadjimichalis (2013), Taibo (2013), Baumgarten (2013), Koutrolikou and Spanou (2013), Kaika and Karliotas (2016), Leontidou (2012), Siatitsa (2014), Diani and Kousis (2014), Gonick (2016), Bosi and Zamponi (2015), Nurra and Azzu (2011), Sevilla-Buitrago (2015), Zavos et al., (2017).

3 There are many other forms of resistance outside urban areas across SE. Among many, see in Spain the movements by migrants in Almeria, the miners' strike in Asturias, the alternative food-sovereignty movement in the Basque Country and the occupation of uncultivated land in Andalusia. In Italy, the movement NO TAV against high-speed train infrastructure in Val di Susa near Turin, the movements against land grabbing in Sardinia and the movements against illegal waste dumping in Campania. In Greece I could mention anti-golf course movements in Crete and Southern Peloponnese, solidarity movements supporting migrants and refugees across Greece, rural strikes by migrants in Epirus and Peloponnese and the large regional social movement against open cast gold mining by the Canadian multinational Eldorado Gold in Halkidiki. See Epochi newspaper, 10 May 2008 and 16 June 2011 and 25 March 2013, in Greek). Velegrakis (2015), Calvário et al. (2016), www.notav.info, D' Alisa et al. (2015), Armiero and D' Alisa (2012), Papataxiarchis (2016a,b).

4 After 1974, the general democratisation of public life in SE was followed by the legalization of the Greek, Spanish and Portuguese Communist Parties. The Greek one, KKE, was divided into two: an orthodox dogmatic faction and a smaller fraction (KKE Interior), which followed Euro-communism and the democratic road to socialism, in line with the Italian and Spanish CPs.

5 The same happened with the strong feminist movement in party politics. After the initial strong push, mainly by Italian feminists, feminism was marginalised and everyday practices returned to traditional worker-based policies. In addition, bitter struggles took place in terms of the autonomy of social movements and often political parties ignored their autonomy or tried to control them, with devastating results.

6 On 20 July 2001, in the international protest against G8 in Genoa, where the slogan was "You G8 – We 8 million", a young protestor, Carlo Giuliani, was shot

to death by Italian carabinieri. It was the first victim of this wave of protest to be followed after seven years by a Greek boy, Alexandros Grigoropoulos in Athens and after ten years by a Turkish boy, Berkin Elvan, in Istanbul.

7 From 1995 until the end of the Games in 2004 dozens of urban grassroots movements emerged in the Athens metropolitan region and in other major cities, partly as a response to projects related to the 2004 Olympic Games and partly as a response to chronic socio-spatial inequalities and injustices in various neighborhoods. These movements were radical in nature, multi-class in social base and quite militant in terms of tactics. These included, among others, movements around the location of specific facilities and their negative effects on the surrounding areas; movements opposing dispossession of public spaces for Olympic use; mobilisations to protect the coastline; movements demanding the implementation of the so-called "green plan" for the games which was suspended due to increased costs; and movements against surveillance and militarization of everyday life in the name of games security. Although sporadic and short-lived in the beginning, in 2003 they were successful in forming a coalition with a committee, a monthly newsletter and a website. Some of those movements remained alive long after the games and have been transformed into radical coalitions for municipal elections (Portaliou, 2007, 2008).

8 In addition, 278 seminars and working groups took place with parallel translation by volunteer interpreters (the famous "Babels"), 104 cultural activities and 35,000 registered participants. Up to 80,000 participated in the final demonstration in major streets of Central Athens.

9 Slogan in the March 2011 Portuguese youth demonstrations.

10 In Napoli, the local campaign is called "Magnammo co pesone", meaning in the local dialect "we will eat the rent".

11 In June 2011, a group of artists occupied the Teatro Valle, Rome's empty oldest theatre. National media paid particular attention and the occupation became one of the country's celebrated anti-austerity actions. During 2011–2013, several other theatres were occupied in Palermo, Catania and Venice. Over the years, these actions have been considered by the activists involved as the parallel in Italy to the indignados movement (www.teatrovalleoccupato.it).

12 See www.transform-network.net; www.euromemo.eu; www.attac.org.

13 See: LIVEWHAT, Living with Hard Times. How Citizens React to Economic Crises and Their Social and Political Consequences, EU 7th Framework Programme (613237) in: www.livewhat.unige.ch/. From newspapers and websites, a very large data base was constructed from which a random sample was drawn. Using a set of codified variables, the next step was the selection from the sample of around 500 active AAOs in each country for further quantitative and qualitative research. Thanks to Maria Kousis for the provision of the report.

14 In the field of social economy and occupied factories-production facilities without bosses, there are only two such factories in operation in Greece, BIO.ME, a chemical company in Thessaloniki, occupied and operated by 55 workers for four years and a factory with 28 workers producing wooden frames in Veria, Northern Greece. They produce for the domestic market with few exports and their experience has been influential on others, such as a textile factory in Veria and some of the small shipyards in Perama (Efimerida ton Syntakton, newspaper, 12 February 2013, in Greek). In the publishing sector there are two successful cooperatives which have arisen since the crisis, the daily newspaper Efimerida ton Syntakton, (The Journal of Journalists) which has the second largest circulation in Greece, and a smaller publishing company for radical books, Colleague's Publishers. Apart from these, there are dozens of small cooperatives in the restaurant and entertainment sector in all major cities.

15 See the work by Greek and foreign social anthropologists among which are Rakopoulos (2015), Cabot (2016), Rozakou (2016), Papataxiarchis (2016a,b,c), and by the social geographer Lafazani (2014, 2017).

16 In Greece, the tradition of alternative economic practices does not have deep roots as in Spain and Italy; it has arisen in the last fifteen years and particularly after 2009. Among the few exceptions are two ecological food-related groups, Peliti since 2001 and Iliosporoi since 2004. They organise native seeds banks, alternative organic agricultural production and various educational and publishing activities. The group Peliti has a national network, with 19 venues in Greece (see www.peliti.gr, and www.iliosporoi.net).

17 The neo-Nazi organisation and political party Golden Dawn made a few attempts to distribute free seasonal food "for Greeks only", introducing openly racist philanthropy.

18 The distinction between refugees and migrants is highly problematic, introduced by international, EU and Greek authorities. Refugees-migrants from Syria could obtain the refugee status and the right to continue their journey while all others could not, despite the fact that all originate from war zones. All these people are refugees and migrants at the same time, so I use the term refugee-migrant.

19 See http://lesvos.w2eu.net/files/2013/03/pikpa.

20 See www.amnesty.org/greece-humanitarian-crisis-refugee/.

21 All Greek and foreign volunteers belong to the radical and anarchist left and have strong anti-Syriza views, expressed in their publications. In private conversations, however, some of them confess that the City Plaza initiative would have been impossible to start and continue until the time of writing if another government had been in power. This issue, however, remains to be seen in the long run due to pressures from the owner and right-wing parties for pushing migrants and volunteers out.

7 Politics of hope or the time of monsters?

Due to the economic crisis, the first two decades of the 21st century changed radically what we once knew as the European Union. Now the "Union" looks more disintegrated than ever, best illustrated in the uneven, multiple and multi-scalar geographies of European integration and imagination. During the economic crisis, the dormant economic and cultural divide between North and South became wider than ever. During the refugee-migrant crisis, another instance of divisive geographical imagination appeared, this time between East and West Europe and the global South. Both crises initiated processes of exclusion and fencing, the first by turning SE societies into "territories of exception" and the second by materially and symbolically building fences to stop the "enemy from the global South". The gradual application of the Maastricht Treaty since the 1990s, the single currency, the Stability and Growth Pact and the Economic Governance of the EU introduced aggressive geographies of welfare deregulation, dispossession, anti-labour rules and regressive taxation, imposing the will of financial capital and rentiers. As European citizens, we face every day the effects of an authoritarian integration that reads like Friedrich Hayek's 20th century fantasy where voters' democratic power should not intervene in the management of the economy and the State's institutions. They must be "independent", that is unaccountable, like the ECB, the Council of Ministers, the Eurogroup and practically all other EU institutions.

Perry Anderson accurately summarises the above discussion:

> The EU is now widely seen for what it has become: an oligarchic structure, riddled with corruption, built on a denial of any sort of popular sovereignty, enforcing a bitter economic regime of privilege for the few and duress for the many.
>
> (*Jacobin,* 23 July 2015)[1]

Thus, it is no surprise that the EU has lost legitimacy at all scales and fuelled political forces hostile not only to this "actually existing" EU, but also to any other integrative alternative. The Brexit vote in the UK and the alarming rise of right-wing, nationalistic parties – some openly neo-Nazi like the Golden Dawn in Greece – have spread scepticism and fear across Europe. The refugee-migrant crisis has awakened xenophobic and racist sentiments, questioning so-called "European solidarity" just as strongly as the economic crisis did before it. The prospect of further fragmentation and enclosure in small nationalistic, uneven, highly competitive and perhaps aggressive territories lies ahead and has been visualised by Umberto Eco who wrote in 1992 that:

> Europe looks integrated but in reality it is more disintegrated than it has been since the eighteenth century. European politicians and bureaucrats do not use history and geography as "*magistrae vitae*" (teachers of life) and cannot visualize the effects should the lid blow off the saucepan.
>
> (Eco, 1992)

The dramatic effects visualised by Eco 25 years ago are already with us, and the rich are those who avoid them, while the poor are those who suffer the pain. European and global elites are jostling for position, and use the crisis – for which they are responsible – as an opportunity to reshape the European and global map to fit their plans.

Is this then the beginning of the end of the EU, as many are asking these days; the end of a project that started 60 years ago and now proves unable to fulfil its promises and looks to be in considerable disarray? I am afraid so, if the situation continues as business as usual, namely imposing austerity, ignoring the structural roots of the Eurozone crisis, creating socio-spatial unevenness and disregarding democracy, because, among other things, the elites don't use history and geography as "magistrae vitae". Then "monsters" may appear in the field of politics if anti-systemic movements and social struggles restrict themselves to the level of identity contradictions as the racist and xenophobic forces do.

Antonio Gramsci, writing from his prison during the end days of the Weimar Republic and before the Second World War, analysed his epoch as being in severe crisis and argued: "...The crisis consists precisely in the fact that the old is dying and the new cannot be born; in this interregnum a great variety of morbid symptoms appear" ["Crisis of Authority" (Selection from the *Prison Notebooks*, 1971: 275–276)]. The second part of the quotation has been popularised as "now is the time of monsters". In this passage, very often quoted these days, Gramsci used the Latin term interregnum to describe a break in the continuity of political regimes, a period during which the normal function of government is suspended. Gramsci introduced these terms to describe the crisis of political hegemony of his own time, but I suggest that they are appropriate to the current European mess and beyond, if we include Trump's USA, the rise of a new Sultan in neighbouring Turkey and the civil war in Syria. We live once again in an interregnum. The question is how to fight the monsters.

What does the analysis in this book tell us about these questions and about the democratic and the anti-capitalist political praxis? It cannot, of course, tell us exactly what to do, but at least it tries to provide a framework for thinking differently, for assessing and judging (*κρίνω,* krino in Greek, the verb that "crisis" comes from) the past facing forwards. The crisis in SE that drove me to write this book has been characterised from the first page as a European/global one, not just a localised consequence of "lazy" people living "beyond their means". My basic argument is that it is not debt but the foundational contradictions of financialisation and uneven geographical development in Europe and the Eurozone that are the roots of the crisis. Undeniably, debt threatens all economies and citizens of Europe (and the world I may add) today and makes immediate relief solutions essential in the extreme cases of Greece and Italy, similar to Germany in the 1950s. Public debt in the Eurozone, however, is used as a political tool to discipline governments and populations, to get governments to implement anti-social policies and to allow capital to gain through dispossessions.

Uneven geographical development is required for capital accumulation because without it, "...capital would surely have stagnated, succumbed to its sclerotic, monopolistic and autocratic tendencies", as Harvey (2014: 161) reminds us. But unevenness across space is not restricted to the needs of capital only. It is combined with the uneven ideological and cultural power that constructs the "Other" as inferior; with the uneven political power of elites and institutions that impose bio-political "exceptional" governance from a distance; with the uneven conditions of wealth, unemployment, education and health; and finally, with the uneven frames of justice that define who counts and which areas count today as subjects of justice.

Political and economic elites ignore these facts on purpose. The elitist, top-down and multi-scalar project of the EU and the Eurozone accepted uneven geographical development under the assumption that compound growth, together with management of the inevitable tensions deriving from the uneven development achieved with structural funds and specialised policies, would permit a smooth evolution. For quite a long time, this assumption did indeed prove to be functional for capital and relatively acceptable to the masses in individual regions and member states. However, major changes in the global division of labour, the collapse of the Soviet Union, the rise of China, the deep transformation of capitalism towards financialisation and rent-seeking activities and the turmoil in 2008–2010 all acted as catalysts to show, once again, the impasse inherent in these assumptions. The wrong architecture of the Eurozone soon proved unable to sustain the new currency union and to handle the euro crisis. While it is a political and spatial matter, it has been left up to those ordoliberal economists promoting austerity to hold the stage. Thus, capitalism is left free to do what it knows best: reproducing itself via "creative" destructions. It survives through the combined operation of spatio-temporal fixes that absorb surpluses, as in northern-central regions of the EU, and by devaluation and destruction of specific areas, as in SE and other places across Europe. Are there victims? Of course there are, as in all wars. But neoliberals have invented new rhetoric and frame their policies as deserved "structural reforms" in the "just" war to "re-establish the confidence of the markets".

Unable or unwilling to face the results of their own practices, European leaders met in March 2017 in Rome to celebrate the EU's 60th "birthday". Among the celebratory speeches, and in the Rome 2017 Declaration, three important concepts, the lack of democracy, austerity and the euro crisis, were absent. It is hardly surprising that political leaders had little self-critical to say about the issue. Instead, in classical Brussels' jargon, the Declaration promised that the Union "will act together, at different paces and intensity where necessary". This is how the elites covered the proposal for a "multi-speed" Europe, launched by Germany, France, Italy and Spain in their exclusive meeting in Versailles.[2] The proposal, known also as "variable geometries", "differentiated integration" and "concentric circles" was around for several years in Brussels and in meetings of experts, but it was opposed by Britain and some other countries. It has been indirectly included in the Maastricht Treaty, the Schengen agreement and finally in the Eurozone. These agreements gave the power to individual states to join policies and treaties selectively on a voluntary basis and, in combination with neoliberalism and financialisation, made clear that real integration and convergence was not an EU target. Uneven development, covered as always with nice words about social and territorial cohesion, was finally accepted by the EU and silently framed as the deserved price for those not willing to join proposed policies.

We don't yet know the exact ingredients of the new proposal except that it openly challenges one of the foundational principles of the then European Common Market, agreed in Rome in 1957, namely that all countries are equal. At that time European integration was a hegemonic project in the sense that the dominant powers promoted their class interests, while looking after the popular masses via social cohesion, redistribution and social welfare. Of course, socio-spatial equality existed only in principle, and in the course of time, capital accumulation and political intervention (e.g. the Maastricht Treaty and the introduction of the euro) made some countries, regions and social groups within them "more equal" than others. Neoliberalism and its German version, ordoliberalism, now dominant in the EU, violently changed the remnants of the old hegemonic project and guided the EU and particularly the Eurozone into crises, ceasarism and austerity. At the beginning of the 21st century, dominant class interests still lead the project but without hegemony. The widespread de-legitimisation of EU policies that came from left anti-systemic movements and parties and, unfortunately, mainly from ultra-right xenophobic political forces is indicative of the present condition.

In the 2017 celebrations in Rome, inequality and uneven development were cynically packed in fancy wrapping and offered as the solution to the EU malaise, as a positive new paradigm supposedly facing forwards while refusing to look backwards, that is without "coming to terms with continuing uneven development in Europe", as Ray Hudson (2017) argues. Some of the Heads of State supporting the idea argued that "multiple speeds already exist in Europe", and Angela Merkel added: "we cannot stop countries wishing to increase their speed of integration". Their cynicism is blind to the fact that the enterprise of European integration, instead of moving at multiple speeds, is at a standstill, or worse, in reverse. Furthermore, it is not accidental that EU leaders and the class interests they represent avoided coming to terms with the existing undemocratic EU structure. In the Rome Declaration, they promised to "promote democratic, effective and transparent decision-making" in clear contradiction to the acceptance of treaties and the practices of EU institutions responsible for applying undemocratic and opaque procedures.

I am aware that the current elitist ruling order in Europe is incapable of dealing effectively with these issues and of restoring confidence and solidarity among sharply divided populations. So who is going to do it? And *should* the European Union even be saved? Answering the second question first, I believe it is worth an effort, although any new founding of another real democratic and just union in Europe appears as a chimera and does not guarantee any success. At least, however, it gives a chance that I personally would not like to miss, since at this juncture, the destruction of the EU would leave free space for monsters to roam in.

Answering the first question and searching for politics of hope in the present European conjuncture, Gramsci's remark on "pessimism of the intellect and optimism of the will" comes to my mind. My analysis so far justifies intellectual pessimism. But in the next moment, I think of the thousands of young people, some of who are anti-capitalist whom I have the privilege to know in person, who flooded the squares and streets of SE, who chose to stay in, or help from abroad, their devastated countries. They all fight in solidarity movements or work in political parties and institutions, and they are all "optimists of the will". Not all of them share the same ideas or the same plan for what to do next. Divisions and fragmentation within movements, particularly of the left, is the rule, but without a minimum basis for understanding and acting together, there is no hope.

SE has been a crucial laboratory for the aggression of ordoliberalism and the socialisation of debt's negative effects. Greece is often framed as the new capitalist development model for the future – a debt colony occupied by rentiers. But Greece and the other countries of SE have also been energetic and promising laboratories of democratic and often anti-capitalist spatialised politics through social movements resisting austerity, forming a mosaic of loosely interconnected transformative movements towards a democratic and sometimes anti-capitalist future. It has also been a laboratory for emerging left-wing political forces, such as Syriza, the Bloco and Izquierda Unida, which preceded the crisis and were strengthened because of it. Others, like Podemos, were born from the movements, particularly the squares' occupations. During 2017, the radical transformation of the British Labour Party under Jeremy Corbyn attracted thousands of new members, mainly young people, and was only 800.000 votes short from winning the election. In France, the radical left alliance "France Insoumise" (Unbowed France) was just 1.7% short of going through to the second round of the presidential elections. Taken all together, and despite their internal contradictions, mistakes and retrogression, important seeds have been planted for politics of hope that allow some optimism of the intellect.

I am interested, therefore, in anti-austerity and anti-capitalist social movements and political forces in SE and in other European countries, some being members of the European Left Party in the EU Parliament while others struggle at their national scale. Obviously I exclude from the politics of hope those populist and extreme right-wing parties and groups that, for various reasons and from completely different ideological standpoints, are nationalistic, xenophobic, racist and strongly against any form of European integration. They exist in all countries and attract substantial popular support. Examples are the National Front in France, the Austrian Freedom Party, the (AfD) Alternative for Germany party and the Netherland's Freedom Party; and then there is Greece's Golden Dawn, which is perhaps more crude and aggressive than the others. In Italy, the M6S or Five Star Movement, a populist anti-establishment and Eurosceptic party, arose as the political outcome of Berlusconi's scandals and the deep crisis of the fragmented Italian left. Equally excluded from politics of hope are neoliberal right-wing political parties and those that belong to what Tariq Ali called the "extremist centre". He meant political forces of the "Third Way" (as per the regional development theories discussed in Chapter 5) that belong to the business-friendly left and to the employee-friendly right, all in support of the EU and in the service of the 1 per cent. Many social-democratic parties went this way through falling in love with neoliberalism: in the UK, Germany, Spain, Italy, Portugal and Greece. Cosmopolitan, politically correct in their speech but ignoring class and socio-spatial inequalities, these political forces were dominant in most countries from the late 1990s and therefore responsible for the current turmoil. After the crisis the extremist centre parties faced "PASOKification", meaning losing their social basis as did PASOK in Greece, although new ones appeared after the crisis like *Ciudadanos* (Citizens) in Spain and The River (*To Potami*) in Greece with restricted popular acceptance. It seems that the new French President Emanuel Macron represents an emerging new version of the extremist centre, now in close collaboration with the global financial capital.

In order to open up ways of thinking that may leave some windows open to politics of hope, I sketch below some ideas based on my SE experience.

I start first by saying that the question of *what kind of Europe "we"*[3] *want cannot be divorced from the question of what kind of people "we" want to be.* To be sure, Europe cannot be changed without changing the dominant political and economic dogma of ordoliberalism. Blaming always the "system", however, without addressing self-critical questions is counterproductive to forming a left, anti-capitalist strategy. Emancipation along class issues is not enough if emancipation along issues of gender, ethnicity, race, religion, age and sexual orientation are not included. Anti-feminist and sexist slogans were frequently heard in demonstrations and square occupations, while often the "dirty" work in solidarity movements has been "women's work". Leftist plans for the future cannot neglect ecological questions and the fact that the everyday behaviour of activists often ignores its ecological footprint. The refugee-migration crisis brought to the surface many left inadequacies that left a vacuum filled by the offensive rise of racist, neo-Nazi ideologies and political practices. The spatialisation of democratic policies at home, the progressive geographical imagination and the relational construction of the "Other", rarely enter a political party's debates. Radical left emancipatory politics cannot move ahead without taking these briefly mentioned issues on board.

Second, an important parameter that requires further attention is the known problem of *how to upscale local bottom-up radical alternatives*. How could we organise together various "militant particularisms", as Raymond Williams once put it? Finally, what do "we" do in normal times, when the dominant capitalist financial system creates cultures of alienating consumerism that cause these radical initiatives to dissolve? These are critical questions for radical activists to debate, among whom there is the known historical split between those who try to search for political alternatives, also at other scales, and those who deny this and concentrate their efforts only on the local scale, which is thus reproduced *ad infinitum*. Massive demonstrations, square occupations and ephemeral solidarity movements are as stimulating as they are deceptive. Or at least they are deceptive if, behind the scenes, no active and lasting radical work is being done at other scales and in institutions. It may be useful to remember at this point Nicos Poulantzas et al. (1980) who argued against the illusion of struggles only at the local level when democratic radical changes in the whole of society, or in the State, are the point at issue. And these struggles could never materialise through processes of direct democracy only while forgetting about representative democracy. And he adds:

> .... what Rosa Luxemburg accused Lenin of was not his neglect of direct democracy but exactly the opposite: that he counted only on direct democracy, which finally ended in the abolition of representative democracy (....) Thus what is at stake for a democratic socialism and transformation of the State is the deepening and broadening of representative democracy with the parallel development of direct democracy and self-managed nuclei at the local level....
>
> (Poulantzas et al., 1980: 161, my translation from Greek)

The movements discussed in Chapter 6 were not fighting for democratic socialism, although some of their practices could be framed so. Poulantzas' remarks are especially vital for those who tend to see large spontaneous gatherings in demonstrations and processes of direct democracy in the squares' open assemblies as isolated events and not as part of the wider social struggle for political and socio-spatial change. The latter needs interventions at other scales, and through other means as well. If "we" don't theorise and practise anti-austerity and anti-capitalist politics simultaneously at various scales, from local to global, and instead propose a loose association of autonomous radical communities – as some participants in the squares' movements have suggested – "we" fail to address the question of uneven development, resulting, as I have highlighted, in socio-spatial inequality and injustice at other scales. Moving beyond sectarianism, however, the struggles should be directed at both the local scale, using direct democracy procedures, and at other scales, through representative democracy, to democratise and reopen all the existing closed and unaccountable EU and Eurozone structures.

Third, and related to the previous point, is the difficult problem of *the relationship between social movements and political parties, or coalitions.* The left tradition in SE on this issue is replete with negative experiences, particularly in Greece and Italy. Although the Syriza victory attracted substantial support from a variety of movements, the coalition government it formed then applied policies contrary to the movements' demands, alienating many left-wing activists. The Spanish case, without a left party in power and without Troika's surveillance, seems to be different, at least at the time of writing. Podemos' electoral growth remains strongly associated with movements and their way of doing politics. The victories of left coalitions in the municipal elections in Barcelona and Madrid, and in other cities such as Valencia and Zaragoza, were outcomes of social movements up-scaling to the metropolitan level. Podemos, as a national political party, is the peculiar and contradictory institutionalisation of some of the participants in the *Indignados* and other social movements that decided to up-scale to enter national and EU politics.[4] Its origin can be found in the manifesto *Mover ficha: convertir la indignación en cambio político* (Move a piece: turn indignation into political change) presented in 2014. Getting around 20 per cent of the national vote, it failed in 2016, in alliance with *Izquierda Unida*, to agree with PSOE to form a coalition government. Here we saw direct interference in Spanish democracy by unaccountable vested class interests, of Spanish and EU origin, to stop a progressive government from taking power. Podemos, and more so the municipal coalitions in Spanish cities, have, however, much to teach other European leftists about how to communicate beyond traditional comfort zones, how to retain a social movement structure in the party's and municipal coalitions' daily routine and how to change traditional governance processes and priorities in order to take 21st-century issues on board.

Fourth, the Eurozone crisis and the authoritarian-undemocratic policies that followed highlighted *the lack of an adequate radical left analysis of the post-political neoliberal Nation-State in the EU and the corresponding issue of popular sovereignty.* A contradiction: the EU was both despised and desired. Although some of those left-wing forces voted against the Maastricht Treaty and were critical of the euro, in successive years, they failed to construct strategies to resist the escalating undemocratic scalar organisation of the EU and the Eurozone. The highly uneven challenge of state sovereignty among member states has been inadequately understood and so has the lack of the euro's political geography. The Eurozone crisis brought to the surface these problems and made urgent a re-conceptualisation of the Nation-State in this post-political, multi-scalar and undemocratic EU. The radical pro-EU left was ill-prepared to face the popular sovereignty issue in a crisis situation, somehow getting lost in its internationalist fantasy of "a Europe of peoples" and in theoretical constructions such as the "hollowing-out" of the Nation-State. In this respect, it has left free space to the anti-EU xenophobic and racist right-wing political forces, like the National Front in France and the (UKIP) UK Independence Party in the UK, while the elites and powerful states, like Germany, manage to secure and progress their class and national sovereignty interests. Equally restrictive for an adequate understanding of these transformations has been the creation of policies focusing only on the local scale, ignoring the national and EU scales. Thus, there is urgent need for radical left rethinking of sovereignty, of citizenship rights to socio-spatial justice and the new role of the Nation-State in the present juncture as part of the wider struggle against neoliberal capitalism and an undemocratic EU regime.

Fifth, from the 1990s onwards, *the organised left often analysed the dynamic of capitalist development with tools and approaches of the post-World War II era.* This is a case of blindness towards the rise of financialisation and a rentier economy, of new political subjectivities and the importance of spatial differentiation at different scales. Resistance to austerity did not arise only from the unions and the working class, narrowly defined as those working in traditional productive sectors, ignoring, for instance, the new precariat. Issues of social reproduction played a more dominant role, shaping new radical demands and fields of struggle, as Harvey (2012) and Fraser (2013) noted. The shift of capital towards rent-seeking profits also transformed the concept of work and class. Although exploitation at the point of production remains central, equal attention should be directed to the circulation and realisation of surplus value in the everyday living spaces of the majority of people. This shift signals on the one hand the increasing multiplicity of political subjects in struggle, and on the other the stronger association between people and places where struggles around social reproduction take place. The latter highlights the importance of socio-spatial unevenness.

Sixth, *if the one side of bio-politics, imposing crude austerity in SE, results in recession, record unemployment, poverty and degradation of the health and educational systems, the other side is the uneven geography of it all.* Socio-spatial contradictions are not the secondary ones resulting from capital's circulation that always becomes a priority after the struggle between capital and labour, as many Marxist economists argue. Place-specific devaluations, as in SE, have proved once more that uneven geographical development among sectors, countries, regions, cities and neighbourhoods is a fertile terrain for profitable accumulation, but also a privileged field for popular struggles and alternative radical policies. The contradictions and conflicts that arise from their uneven management highlight more clearly the socio-spatial inequalities and injustices and make people understand their exploited situation better than the abstract categories used by the jargon of the traditional left. Thus, there is an urgent need for the left in SE, and perhaps globally, to understand the importance of spatialised politics, so evident in most of the social movements discussed in this book and to listen to those activists engaged with them, learn from their experience and appreciate their wider social recognition, as in the case of the two women mayors in Spain.

To do this, the organised left faces two traditional difficulties. On the one hand, if it recognises the importance of space in politics, this is reduced to physical space only, forgetting representation of space, spatial imaginations/representations and the ecological parameters. Space and uneven geographical development are not "external environments". They constitute an indispensable part of conflicting social relations. On the other, the organised left is often the prisoner of statism and sits uncomfortably with social movement procedures that are more suitable for dealing with socio-spatial contradictions. Here again the problem of scale reappears, but as is known, every problem has its own appropriate scale where solutions must be sought.

My final point is that *the optimism of the will is not enough to get radical left politics applied.* The previous points need to be taken on board and a coherent strategy is essential, together with good knowledge of enemies and continuous support from social movements and the masses. In short, it needs hegemony, not just a change of government. The case of Syriza-led coalition government perfectly illustrates the omission of these points. Although its two electoral victories were major breakthroughs that shocked the Greek and European establishment, its subsequent policies and decisions disappointed its leftist audience in Greece and globally. The manipulation of the OXI (NO) vote in the 2015 referendum and the signing of the third Memorandum were major negative turning points, despite the lack of socially acceptable alternatives at that period.[5] Many leftists in Greece and abroad had high expectations regarding radical changes from a Syriza government, despite the fact that the coalition government took office under the crisis conditions, with a destroyed State apparatus and highly indebted economy and the tough surveillance of the Troika. However, in non-revolutionary periods, major radical changes do not occur in a short time period. These points do not mean that as a radical leftist I feel satisfied with what is going on in my country; quite the contrary.

There is by now a booming left-wing critique against Syriza, even from activists and party members inside the party, focusing on several issues. It seems that there is a lack of understanding among Syriza's leadership that the so-called "recovery plan" in the third Memorandum cannot lead to exit from the crisis. The anaemic growth of GDP since 2015 and the unrealistic 3.5 per cent of GDP primary surplus every year will simply reproduce recession and austerity. On top of this, the debt problem remains unresolved. On the everyday life front, where a leftist strategy is weighed against the realities of the crisis, many fronts are moving along the wrong path despite some important positive steps in public health and education, social protection and progressive changes in planning law. For example, pension cuts and uncontrolled precarity continue; there is no relation with social movements, and some cabinet members often show arrogance towards the limited social protest; younger radicals working in institutions are lost in the bureaucracy of the destroyed state apparatus; the promises of a democratic and transparent transformation of State institutions remains unfulfilled; and there is no sign of any will to fight against dispossession of public land and public utility companies. On the contrary, in several cases, Syriza submitted to pressures from vested interests to make real estate investments against the interests of local social movements.

There is also the debate around a voluntary "Grexit" from the Eurozone. I would not directly engage in this debate, but I have noted the vague description of Grexit's outcomes at multiple scales and particularly in the everyday life of already exhausted people. However, an exit decision requires good timing and preparation, which does not exist, and popular legitimacy, i.e. a referendum, a fact often conveniently denied by Grexit advocates.

Judging the Syriza experience from a left-wing point of view is perhaps premature, although its policies have shown the limits of a radical left strategy against the EU and the Eurozone establishment. The Portuguese experience in building a government coalition is different and until the time of writing, seems interesting, bearing in mind that Portugal is not anymore under the Troika. The PCP, the Greens and the radical left Bloco supported the moderate Socialist Party under a new leader, António Costa, to form a government with the aim of rolling back austerity. They agreed a minimum programme consisting of three points: raising the minimum wage, lifting a freeze on pensions and cancelling pay cuts to civil servants. Lisbon, once Merkel's foster child, went to the moderate left through *geringonça* – that roughly translates as "contraption" – a historical compromise à la Portuguese, in which the radical left parties can step aside and criticise the government every time they feel it necessary. However, Portugal manages to borrow at low rates from international markets in a completely different economic context to Greece and has more freedom for anti-austerity policies, although Brussels and Berlin often intervene through the new EU Economic Order to control its budget. Contrary to Syriza, however, left-leaning supporters are in favour of the government and its politics.

Despite limits and contradictions, the radical left and particularly the anti-capitalist left is the only political force capable of re-politicising and re-democratising politics. A new European radical left is required that starts from the bottom up in each country, and at the EU scale, and fights on five combined fronts: to reverse policies that reproduce austerity and recession at home and in the EU; to restructure the debt; to reorganise the euro architecture and the function of the ECB; to challenge policies fuelling uneven development within each country and across the EU; and to radically democratise the institutions at home and within the EU. Can we do it? I think "we" can, but not without a rupture with the national elites and disobedience to the European establishment. No country can do it alone, focusing only on domestic policies, as some leftists argue, although a re-conceptualisation of the role of Nation-State and popular sovereignty is essential. A pan-European radical left solidarity movement is highly needed, capable of mobilising the millions of citizens deprived by austerity and undemocratic governance. Is this realistic or an unreachable utopia? The answer lies in political, class and socio-spatial struggles, not in theories; a utopia – another spatial term – that guides the radical practices, both successful and unsuccessful, of all left-wing struggles. Because as Eduardo Galeano (1995) writes:

> Utopia, is on the horizon, says Fernando Birri. I go two steps closer, she moves two steps away. I walk ten steps and the horizon runs ten steps ahead. ... What good is utopia? Here's what: it's good for making us keep walking forward.

Notes

1 See www.jacobinmag.com/2015/07/tspiras-syriza-euro-perry-anderson/.

2 The arrogance and undemocratic process that symbolise the Versailles meeting in March 2017 highlights the European interregnum. The proposal that will affect millions of people if applied was taken by the temporary Italian PM Gentiloni, the Spanish PM Rajoy, who leads a government that could be overthrown at any moment, the French president Hollande, enjoying the lowest popularity ever and while he was about to leave office in a few months and the German Chancellor, Angela Merkel, whose dominance in German politics is for the first time being seriously challenged.

3 In what follows I use "we", in quotation marks. I mean a contested coming together of people belonging to a "Radical Left Multitude", not necessarily of the same party or in the same country.

4 Podemos has MPs in both the Spanish and European parliaments. Other *Indignados* joined the right-wing party of Ciudadanos (citizens), supporting PP.

5 After the defeat and acceptance of a third Memorandum in June 2015, several comrades left Syriza and formed the Popular Unity Party, which was unable to enter the Parliament in the September 2015 elections. They strongly support exiting from the Eurozone. Many other Syriza members left the party as well and organised small discussion groups here and there.

References

Abellàn, J., Sequera, J., Janoschka, M. (2012) "Occupying the #Hotelmadrid: a laboratoty of urban resistance", *Social Movements Studies*, 11(3–4): 320–326.

Accornero, G., Ramos Pinto, P. (2015) "'Mild Mannered'? Protest and mobilisation in Portugal under austerity, 2010–2013", *West European Politics*, 38(3): 491–515.

Agamben, G. (1998) *Homo Sacer: Sovereign Power and Bare Life*, Stanford: Stanford University Press.

Aglietta, M. (2000) "Shareholder value and corporate governance: some tricky questions", *Economy and Society*, 29(1): 146–159.

Akin, O., García Montalvo, J., García Villar, J., Peydró, J-L., Raya, J.M. (2014) "The real estate and credit bubble: evidence from Spain", www.researchgate.net/publication/265013899.

Albertos, J.M., Sánchez, J.L. (eds) (2014) *Geografía de la crisis económica en España*, València: Publicacions de la Universitat de València.

Alesina, A., Ardagna, S. (2009) "Large changes in fiscal policy: taxes versus spending", Working Paper 15438. www.nber.org/papers/w15438.

Ali, T. (1999) "Springtime for NATO", *New Left Review*, 234: pp. 62–75.

Amin, A. (2003) "Industrial districts", in: Barnes, T.J., Sheppard, E. (eds) *A Companion to Economic Geography,* Blackwell: Oxford, pp. 149–168.

Amin, A., Thrift, N. (2005) "What's left? Just the future", *Antipode,* 37(2): 220–238.

Ancelovici, M., Dufour, P., Héloïse, N. (eds) (2016) *Street Politics in the Age of Austerity. From the Indignados to Occupy*, Amsterdam: Amsterdam University Press.

Andreotti, A., Mingione, E. (2014) "Local welfare systems in Europe and the economic crisis", *European Urban and Regional Studies*, 23(3): 231–251.

Anthias, F., Lazaridis, G. (2000) "Introduction: women on the move in southern Europe", in: Anthias, F., Lazaridis, G. (eds) *Gender and Migration in Southern Europe*, Oxford: Berg, pp. 1–14.

ARMAL (Agenzia Regional Marche Lavoro). (2003) *Economia e Territorio. Il distretto calzaturiero fermano-maceratese*, Ancona.

Armiero, M., D'Alisa, G. (2012) "Rights of resistance: the garbage struggles for environmental justice in Campania, Italy", *Capitalism Nature Socialism,* 23(4): 52–68.

Arrighi, G. (ed) (1985) *Semiperipheral Development. The Politics of Southern Europe in the Twentieth Century*, London: Sage.

Arrighi, G. (1990) "The Developmentalist illusion: a reconceptualization of the semiperiphery", in: Martin, W.G. (ed) *Semiperiheral States in the World Economy*, Westport, CT: Greenwood Press, pp. 11–42.

Arrighi, G. (1994) *The Long Twentieth Century: Money, Power and the Origins of Our Times*, London: Verso.

Asheim, B. (1996) "Industrial districts as 'learning regions': a condition for prosperity?", *European Planning Studies,* (4)4: 379–400.

Athanasiou, A. (2012) *The Crisis as "a State of Exception". Critiques and Resistance*, Athens: Savalas (in Greek).

Aziz, J. (2015) "The trouble with ordoliberalism", www.pieria.co.uk/articles/the_trouble_with_ordoliberalism

Bagnasco, A. (1977) *Tre Italie. La problematica territoriale dello sviluppo italiano,* Bologna: il Mulino.

Balabanidis, D., Patatouka, E., Siatitsa, D. (2013) "The right to housing during the crisis in Greece", *Geographies*, 22: 31–42 (in Greek).

Baldwin-Edwards, M., Arango, J. (eds) (1999) *Immigrants and the Informal Economy in Southern Europe*, London: Franc Cass.

Barro, R., Sala-i-Martin, X. (1995) *Economic Growth,* New York: Mcgraw Hill.

Baumgarten, Br. (2013) *"Geração à Rassa* and beyond: mobilisations in Portugal after 12 March 2011", *Current Sociology*, 61(4): 457–473.

Becattini, G., Bellandi, M., Dei Ottati, G., Sforzi, F. (eds) (2003) *From Industrial Districts to Local Development*, Cheltenham: E. Elgar.

Bellina, B. (2013) "Germany in times of crisis: passive revolution, struggle over hegemony and new nationalism", *Geografiska Annaler*, 95(3): 275–285.

Bellofiore, R. (2013) "Two or three things I know about her: Europe in the global crisis and heterodox economics", *Cambridge Journal of Economics*, 37: 497–512.

Belso-Martínez, J.A. (2010) "International outsourcing and partner location in the Spanish footwear sector: and analysis based in industrial district SMEs", *European Urban and Regional Studies*, 17(1): 65–82.

Benach, N. (2015) "Contested discourses of austerity in the urban margins (a vision from Barcelona)", in: Echardt, F., Ruiz-Sánchez, J. (eds) *City of Crisis. The Multiple Contestation of Southern European Cities*, Bielefeld: Transcript Verlag, pp. 31–49, 71–85.

Beramendi, P. (2012) *The Political Geography of Inequality: Regions and Redistribution*, Cambridge: Cambridge University Press.

Bergfeld, M. (2014) *Portugal, 40 years after the Revolution 1974–2014*, mdbergfeld.com/2014/04/portugal-40-years-after-the-revolution-ebook/.

Bertoncin, M., Marini, D., Pase, A. (2009) *Frontiere mobili: delocalizzazione e internazionalizzazione dei territori produttivi veneti*, Venezia: Marsilio.

Bianchi, P., Bellini, N. (1991) "Public policies for local network of innovations", *Research Policy*, 12: 3, 487–497.

Blim, M. (1989) "Prima e dopo lo sviluppo. Monte San Giusto dall'Unità ad oggi", in: Anselmi, S. (ed) *L'industria calzaturiera marchigiana. Dalla manifattura alla fabrica,* Fermo, pp. 38–49.

Blim, M. (1990) *Made in Italy: Small Scale Industrialization and Its Consequences*, New York: Preager.

Boschma, R.A. (2004) "The competitiveness of regions from an evolutionary perspective", *Regional Studies*, 38: 993–1006.

Boschma, R.A., Frenken, K. (2006) "Why is economic geography not an evolutionary science? Towards an evolutionary economic geography", *Journal of Economic Geography*, 6: 273–302.

Bosi, L., Zamponi, L. (2015) "Direct social actions and economic crises: the relationship between forms of action and socio-economic context in Italy", *PArtecipazione e COnflitto*, 8(2): 367–391.

Bourdieu, P. (1980) "Le capital social", *Actes de la Recherche en Sciences Social*, 3(2–3): 123–134.

Bourdieu, P. (1990) *The Logic of Practice*, Stanford: Stanford University Press.

Boyer, R. (2000) "Is a finance-led growth regime a viable alternative to Fordism? A preliminary analysis", *Economy and Society*, 29(1): 111–145.

Braudel, F. (1966) *La Méditerranée et le Monde Méditerranéan à l' Epoque de Philippe II*, vol. 1 et 2, Paris: Librairie A. Colin.

Brenner, N. (1997) "Global, fragmented, hierarchical: Henri Lefebvre's geographies of globalization", *Public Culture*, 10: 137–169.

Bruff, I. (2011) "What about the elephant in the room? Varieties of capitalism, varieties in capitalism", *New Political Economy*, 16(4): 481–500.

Burroni, L. (2016) *Capitalismi a confronto. Istituzioni e regolazione dell'economia nei paesi europei,* Bologna: il Mulino.

Buzzati, St., Pasquato, Ch. (2009) "Tra successo e fallimento. La sfaccettature della territorialità nel distretto dello SportSystem di Montebelluna", in: Bertoncin, M., Marini, D., Pase, A. (eds) *Frontiere Mobili. Delocalizzazione e internationalizzazione dei territori produttivi veneti,* Venezia: Marcilio, pp. 47–70.

Cabot, H. (2016) "Contagious" solidarity: reconfiguring care and citizenship in Greece's social clinics", *Social Anthropology*, 24(2): 152–166.

Cadelli, M. (2016) "Le néoliberalism est un fascism", *Le Soir*, 3 March 2016.

Calvário, R., Velegrakis, G., Kaika, M. (2016) "The political ecology of austerity: an analysis of socio-environmental conflict under crisis in Greece", *Capitalism Nature Socialism*, doi:10.1080/10455752.2016.1260147.

Cánovas, A.P., Riquelme Perea, J. (2007) "La condición inmigrante de los nuevos trabajadores rurales", *Revista Española de Estudios Agrosociales y Pesqueros,* 211: 189–238.

Capecchi, V., Pesce, A. (1983) "Se la diversità è un valore", *Inchiesta*, 13: 59–60.

Carvalho, L. (2017) "Life after the squares: reflections on the consequences of the occupy movements" (intervention on Rossio), *Social Movements Studies*, 16(1): 119–151.

Castells, M. (1983) *The City and the Grassroots. A Cross-Cultural Theory of Urban Social Movements*, London: Arnold.

Castells, M. (2016) *Networks of Outrage and Hope: Social Movements in the Internet Age*, London: Polity.

Castro, M., Garcia, B., Vatavali, F., Zifou, M. (2013) "Ultra-neoliberal urban development in Spain and Greece. The case of Port Vell in Barcelona and Hellinikon Airport in Athens", *Geographies*, 22: 14–29 (in Greek).

Cecchi, E., Seassaro, A., Simonelli, G., Sorlini, C. (1978) *Centri Sociali autogestiti e circoli giovanili*, Milano: Fetrinelli.

Cervellati, P., Scannavini, R. (1973) *Bologna. Politica e metodologia del restauro nei centri storici*, Bologna: Il Mulino.

Charnock, G., Purcell, Th., Ribera-Fumaz, R. (2014) *The Limits to Capital in Spain: Crisis and Revolt in the European South*, London: Palgrave MacMillan.

Chomsky, N. (1999) *The New Military Humanism: Lesson from Kosovo*, London: Pluto Press.

Clark, G. (2014) "The geography of the Euro Crisis: the ECB, its institutional form, functions and performance". Paper presented at AAG, Florida, April 2014 (available from the author).

Clark, J., Jones, A. (2008) "The spatialities of Europeanisation: territory, government and power in Europe", *Transactions of the Institute of British Geographers*, NS 33, 300–318c.

Cómo cocinar una revolutión no violenta, http://takethesquare.net/es/2011/08/18/como-cocinar-una-revolution-no-violenta/.

Compaña Quién Debe a Quién. (ed) (2011) *Vivir en deudocracia. Iban un portugués, un irlandés, un griego y un español....*Barcelona: Icaria.

Conill, J., Castells, M., Cardenas, A., Servon, L. (2012) "Beyond the crisis: alternative economic practices in Catalonia", in: Castells M., Caraçao, J., Cardoso, G. (eds) *After-Math. The Culture of Economic Crises*, Oxford: Oxford University Press, pp. 211–248.

Cooke, P. Morgan, K. (1998) *The Associational Economy: Firms, Regions and Innovation*, Oxford: Oxford University Press.

Coq-Huelva, D. (2013) "Urbanisation and financialisation in the context of a rescaling state: the case of Spain", *Antipode*, 45(5): 1213–1231.

Credit Suisse. (2016) *Global Wealth Data Book*, www.credit-suisse.com/articles/2016/the-global-wealth-report-2016.ht

Crestanello, P. (1999) *L' industria veneta di abbigliamento: internazionalizzazione produttiva e piccole imprese di sub-fornitura*, Milano: F.Angeli.

Croce, B. (1925/1970 transl) *History of the Kingdom of Naples*, Chicago: University of Chicago Press.

Cutrini, E. (2011) "Moving Eastwards while remaining embedded: the case of the Marche footwear district, Italy", *European Planning Studies*, 19(6): 991–1014.

D' Alisa, G., Forno, F., Maurano, S. (2015) "Grassroots (economic) activism in times of crisis", *PArtecipazione e COnflitto*, 8(2): 328–342.

Davis, M. (2011) "Spring confronts Winder", *New Left Review*, 72: 5–15.

De Cesaris, W. (2009) "The speculative real-estate bubble and the securitisation scandal in Italy", www.transform-network.net/yearbook/journal-052009/9.

Del Monte, A., Giannola, A. (1978) *Il Mezzogiorno nell' economia italiana*, Bologna: Il Mulino.

Della Porta, D. (2015) *Social Movements in Times of Austerity: Bringing Capitalism Back into Protest Analysis*, London: Polity.

Della Porta, D., Mattoni, A. (eds) (2014) *Spreading Protest. Social Movements in Times of Crisis*, Colchester: ECPR Press.

Diani, M., Kousis, A. (2014) "The duality of claims and events: the Greek campaign against the Troika's Memoranda and austerity, 2010–2012", *Mobilization: An International Quarterly,* 19(4): 387–404.

Dicken, P. (2015) *Global Shift. Mapping the Changing Contours of the World Economy*, London: Sage (seventh edition).

Dijstelbloem, H., Meijer, A. (eds) (2011) *Migration and the New Technological Borders of Europe*, London: Palgrave MacMillan.

Dinoto, I. (2013) "Housing crisis in Italy", *Geographies*, 22: 43–44.

Dorling, D. (2010) *Injustice. Why Social Inequality Persists*, Bristol: The Polity Press.

Douzinas, C. (2013) "Athens rising", *European Urban and Regional Studies,* 20(1): 134–138.

Dunford, M., Dunford, R., Barbu, M., Weidong, L. (2013) "Globalisation, cost competitiveness and international trade: the evolution of the Italian textile and clothing industries and the growth of trade with China", *European Urban and Regional Studies,* 21: 1–25.

Dunford, M., Yeung, G. (2009) "Port-Industrial Complexes", in: Kitchen, R., Thrift, N. (eds) *International Encyclopedia of Human Geography*, Regional Development Section, Elsevier, www.elsevier.com/HUGY/Dunford_Yeung.

ECB (2015) *Real estate markets and macroprudential policy in Europe*, ECB Working Paper 1796, May 2015.

Eco, U. (1992) "Qual è il costo di un impero fallito?", *L' Espresso*, 6 September.

Eizaguirre, S., Pradel, M., Garcia, M. (2017) "Citizenship practices and democratic governance: "Barcelona en Comú" as un urban citizenship confluence promoting a new policy agenda", *Citizenship Studies*, 21(4): 425–439.

Ekers, M., Hart, G., Kipfer, St., Loftus, A. (eds) (2013) *Gramsci. Space, Nature, Politics*, Chichester: Wiley-Blackwell.

Eurispes (2001) *Rapporto Italia*, Roma: Ufficio Stampa Eurispes.

Europe 2020 (2010) A European Strategy for Smart, Sustainable and Inclusive Growth, Brussels, www.ec.europa.eu/eu2020/pdf/COMPLETE%

Fazi, Th. (2017) "Public debt in the Eurozone: a political problem", speech given at the conference *"How to Deal with Public Debt?-Lesson Learned and Policies Ahead"*, 7 March, *GUE/NGL*, European Parliament, www.socialeurope.eu/2017/03/public-debt-eurozone-political-problem-financial-one/.

Featherstone, D. (2012) *Solidarity: Hidden Histories and Geographies of Internationalism*, London: Zed Books.

Featherstone, D. (2013) "Gramsci in action": space, politics and the making of solidarities", in: Ekers, M., Hart, G., Kipfer, St., Loftus, A. (eds) *Gramsci. Space, Nature, Politics,* Chichester: Wiley-Blackwell, pp. 65–82.

Fernández-Savater, A., Flesher Fominaya, C. (ed) (2017) "Life after the squares: reflections on the consequences of the occupy movements", *Social Movement Studies*, 16(1): 119–151 (with contributions by Carvalho, L., Hoda Elsadda, G., El-Tamami, W., Horrillo, P., Nanclares, S., Stavrides, S.).

Flassbeck, H. (2010) "Avis de tempête sur l'Union monétaire européenne", *Le Monde,* 5 March 2010.

Fondazione Nordest. (2003) *La de-localizzazione produtiva all estero nel settore moda: il Caso Vicenza*, Venezia (mimeo).

Foster, J.B. (2007) "The financialization of capitalism", *Monthly Review*, 58(11).

Fraser, N. (1995) "From redistribution to recognition? Dilemmas of justice in a 'post-socialist' age", *New Left Review*, 212: 68–93.

Fraser, N. (2006) "Reframing justice in a globalizing world", *New Left Review*, 36, Nov-Dec: 69–88.

Fraser, N. (2008) *Scales of Justice. Reimagining Political Space in a Globalizing World*, Cambridge: Polity.

Fraser, N. (2013) "A triple movement?", *New Left Review*, 81, May-June: 119–132.

Fraser, N. (2016) "Crises of care: the contradictions of social reproduction in the era of financial capitalism", *The Annual Nicos Poulantzas Lecture*, Nicos Poulantzas Institute, Athens, 7 December 16. http://poulantzas.gr/rethinking-greece-nancy-fraser/.

Fujita, M., Krugman, P. (2004) "The new economic geography: past, present and future", Papers in *Regional Science*, 83: 139–164.

Galeano, E. (1995) *Walking Words*, New York: Gorton & Company, (translated by Mark Fried).

Galeano, E. (2004) "Interview", in: Barsamian, D. (ed) *Louder than Bombs: Interviews from the Progressive Magazine*, New York: South End Press, p. 146.

Gambarotto, F., Solari, St. (2015) "The peripheralization of Southern European Capitalism with the EMU", www.siecon.org/online/wp-content/uploads/2013/09/Gambarotto-Solari.

García, M. (2010) "The breakdown of the Spanish urban growth model: social and territorial effects of the global crisis", *International Journal of Urban and Regional Research*, 34(4): 967–980.

Garofoli, G. (1983) *Industrializazione diffuza in Lombardia*, IPER/F. Milano: Angeli.

Gialis, S., Herrod, A. (2014) "Of steel and strawberries: Greek workers struggle against informal and flexible working arrangements during the crisis", *Geoforum*, 57: 138–149.

Gialis, St., Leontidou, L. (2016) "Antinomies of flexibilization and atypical employment in Mediterranean Europe: Greek, Italian and Spanish regions during the crisis", *European Urban and Regional Studies*, 23(4): 716–733.

Gibelli, M.-C. (2015) "Urban crisis or urban decay? Italian cities facing the effects of a long wave towards privatization of urban policies and planning", in: Echardt, F., Ruiz-Sánchez, J. (eds) *City of Crisis. The Multiple Contestation of Southern European Cities*, Bielefeld: Transcript Verlag, pp. 89–108.

Giner, S., Sevilla, E. (1984) "Spain: from corporatism to corporatism", in: Williams, A. (ed) *Southern Europe Transformed*, London: Harper and Raw, pp. 113–144.

Golemis, H. (2010) "Can PIGS Fly?" *Transform! European Journal for Alternative Thinking and Political Dialogue*, 6: 129–136.

Gonick, S. (2016) "Indignation and inclusion: activism, difference, and emergent urban politics in postcrash Madrid", *Environment and Planning D: Society and Space*, 34(2): 209–226.

Gonzáles-Diaz, B., Gandoy, R. (2005) "Understanding offshoring: has Spain been an offshoring location in the 1990s?" www.ub.edu/jei/papers/GONZALEZ-GANDOY.pdf.

Grabher, G. (2009) "Yet another turn? The evolutionary project in economic geography", *Economic Geography*, 85(2): 119–127.

Grazioli, M. (2017) "From citizens to *citadins*? Rethinking right to the city inside housing squats in Rome, Italy", *Citizenship Studies*, 21(4): 393–408.

Greber, D. (2011) *Debt. The first 5,000 years*, New York: Mellvile House Publ.

Gregory, D. (1995) "Imaginative geographies", *Progress in Human Geography*, 19: 447–485.

Gregory, D. (2004) *The Colonial Present*, Oxford: Blackwell.

Guldi, J., Armitage, D. (2014) *The History Manifesto*, Cambridge: Cambridge University Press.

Hadjimichalis, C. (1987) *Uneven Development and Regionalism: State, Territory and Class in Southern Europe*, London: Croom Helm.

Hadjimichalis, C. (2006) "The end of third Italy as we knew it?", *Antipode*, 38(1): 82–106.

Hadjimichalis, C. (2010) "The Greek economic crisis and its geography: from R. Kaplan's geographical determinism to uneven geographical development", *Human Geography: A New Radical Journal*, 3(3): 89–100.

Hadjimichalis, C. (2011) "Uneven geographical development and socio-spatial justice and solidarity: European regions after the 2009 financial crisis", *European Urban and Regional Studies,* 48(3): 254–274.

Hadjimichalis, C. (2013) "Luchas urbanas y construcción de redes de solidaridad en Atenas durante la crisis", *Urban,* NS06: 79–97.

Hadjimichalis, C. (2014) "Crisis and land dispossession in Greece as part of the global 'land fever'", *City: Analysis of Urban Trends, Culture, Theory, Policy, Action*, 18(4–5): 502–508, doi:10.1080/13604813.2014.939470.

Hadjimichalis, C. (2014) *Debt Crisis and Land Dispossession*, Athens: ΚΨΜ Publications (in Greek). See also the German translation: *Schuldenkrise and Landraub in Griechenland*, Münster: Westfälishe Dampfboot (2016).

Hadjimichalis, C., Hudson, R. (2004) "Networks, regional development and democratic control", in: *Proceedings of Seminars of the Aegean, Naxos 2003*, Athens-Thessaloniki, pp. 123–139.

Hadjimichalis, C., Hudson, R. (2007) "Rethinking local and regional development: implications for radical political practice in Europe", *European Urban and Regional Studies,* 14(2): 99–113.

Hadjimichalis, C., Hudson, R. (2014) "Contemporary crisis across Europe and the crisis of regional development theories", *Regional Studies*, 48(1): 208–218.

Hadjimichalis, C., Papamichos, N. (1990) "Local development in Southern Europe: towards a new mythology", *Antipode*, 22(3): 181–200.

Hadjimichalis, C., Sadler, D. (eds) (1995) *Europe at the Margins. New Mosaics of Inequality*, Chichester: Wiley.

Hadjimichalis, C., Vaiou, D. (1990) "Whose flexibility? The politics of informalisation in Southern Europe", *Capital and Class,* 42, Winter: 79–106.

Hadjimichalis, C., Vaiou, D. (2004) "Local" illustrations for "International" geographical theory", in: Simonsen, K., Bœrenholdt, J.O. (eds) *Space Odysseys*, Aldershot: Ashgate, pp. 171–182.

Harrison, B. (1994) "The Italian industrial districts and the crisis of the cooperative form: part I", *EPS,* vol. 2:1, pp. 3–22 and part II, *EPS,* vol. 2:2, pp. 159–174.

Harvey, D. (1973) *Social Justice and the City*, Baltimore: Johns Hopkins University Press.

Harvey, D. (1982) *The Limits to Capital*, Oxford: B. Blackwell.

Harvey, D. (2003) *The New Imperialism,* Oxford: Oxford University Press.

Harvey, D. (2009) "The right to the Just City", in: Marcuse, P., Connolly, J., Novy, J., Olivo, I. (eds) *Searching for the Just City*, London: Routledge, pp. 40–51.

Harvey, D. (2010) *The Enigma of Capital,* London: Profile Books.

Harvey, D. (2011) "Crises, geographic disruptions and the uneven development of political responses", *Economic Geography,* 87(1): 1–22.

Harvey, D. (2012) *Rebel Cities: From the Right to the City to the Urban Revolution*, London: Verso.

Harvey, D. (2014) *Seventeen Contradictions and the End of Capitalism*, London: Profile Books.

Hay, C. (1999) "Crisis and the structural transformation of the state: interrogating the process of change", *British Journal of Politics and International Relations*, 1(3): 317–344.

Healey, P. (1997) *Collaborative Planning: Shaping Place in Fragmented Societies*, London: Macmillan.

Hodgson, G.M. (2005) "Generalizing Darwinism to social evolution: some early attempts", *Journal of Economic Issues*, 39: 899–914.

Horrilo, P., Nanclares, S. (2017) "Life after the squares: reflections on the consequences of the occupy movements" (intervention on Puerta del Sol), *Social Movements Studies*, 16(1): 119–151.

Hudson, R. (1999) "The learning economy, the learning firm and the learning region: a sympathetic critique of the limits to learning", *European Urban and Regional Studies*, 6(1): 59–72.

Hudson, R. (2001) "Regional development, flows of value and Governance in an Enlarged Europe", *Regional Regeneration and Development Studies*, University of Durham (mimeo).

Hudson, R. (2005) *Economic Geographies-Circuits, Flows and Spaces*, London: Sage.

Hudson, R. (2007) "Regions and regional uneven development forever? Some reflective comments upon theory and practice", *Regional Studies*, 14(9): 1149–1160.

Hudson, M. (2010) "The transition from industrial capitalism to a financialized bubble economy", Working Paper 627, Levy Economics Institute, www.levyinstutute.org/pubs/wp_627.

Hudson, R. (2017) "Facing forwards, looking backwards: coming to terms with continuing uneven development in Europe", *European Urban and Regional Studies*, 24(2): 138–141.

Hudson, R., Lewis, J. (1985) *Uneven Development in Southern Europe*, London: Methuen.

Il Sole 24ore (2002) "Carpi un distretto all cinese", 19 July 2002.

INE/GSEE (2010) *The Greek Economy and Employment*, Annual Report, Athens (in Greek): Greek Labour Institute.

IOBE (2015) *The Importance, Obstacles and Future of the Construction Sector in Greece*, Athens (in Greek): IOBE.

Jabko, N. (2010) "The hidden face of the euro", *Journal of European Public Policy*, 17(3): 318–334.

Jessop, B. (2005a) "Gramsci as a spatial theorist", *Critical Review of International Social and Political Philosophy*, 8(4): 561–574.

Jessop, B. (2005b) "The political economy of scale and European governance", *Tijdschrift voor Economische en Sociale Geographie*, 96(2): 225–230.

Jessop, B. (2011) "Rethinking the diversity and variability of capitalism. On variegated capitalism in the world market", in: Lane, C., Wood, G. (eds) *Capitalist Diversity and Diversity within Capitalism*, London: Routledge, pp. 211–237.

Kaejane, G. (2011) "Seven key words for the Spanish experience, Puerta del Sol, M15", www.edu-factory.org/wp/spanishrevolution (accessed 15 December 2012).

Kaika, M. (2012) "The economic crisis seen from the everyday: Europe's nouveau poor and the global affective implications of a "local" debt crisis", *City*, 16(4): 422–429.

Kaika, M., Karaliotas, L. (2016) "The spatialisation of democratic politics: insights from the Indignant Squares", *European Urban and Regional Studies*, 22(4): 556–570.

Kalogeresis, A., Labrianidis, L. (2008) "Delocalization and development in Europe: conceptual issues and empirical findings", in Labrianidis, L. (ed) *The Moving Frontier. The Changing Geography of Production in Labour-Intensive Industries*, Aldershot: Ashgate, pp. 23–58.

Kaplan, R. (1994) *Balkan Ghosts: A Journey through History*, New York: Vintage Books.

Kaplan, R. (2009) "The revenge of geography", *Foreign Policy*, 179: 96–105.

Kaplan, R. (2010) "For Greece's economy, geography was destiny", *The New York Times*, 25 April 2010.

Karamesini, M. (2013) "Structural crisis and adjustment in Greece: social regression and the challenge to gender equality", in: Karamesini, M., Rubery, J. (eds) *The Economic Crisis and the Future of Gender Equality*, London: Routledge, pp. 165–185.

Karamesini, M., Rubery, J. (eds) (2013) *The Economic Crisis and the Future of Gender Equality*, London: Routledge.

Kearns, G. (2009) "Mackinder Redux", *Human Geography*, 2(2): 44–48.

King, R. (2000) "Southern Europe in the changing global map of migration", in: King, R., Lazaridis, G., Tsardanidis, Ch. (eds) *Eldorado or Fortress? Migration in Southern Europe*, London: MacMillan, pp. 1–26.

Kousis, M. (2016) "The spatial dimension of the Greek protest campaign against the Troika's Memoranda and austerity, 2010–1013", in: Ancelovici, M., Dufour, P., Nez, H. (eds) *Street Politics in the Age of Austerity*, Amsterdam: Amsterdam University Press, pp. 147–174.

Kousis, M., Kalogeraki, St., Papadaki, M., Loukakis, A., Velonaki, M. (2016, forthcoming) "Alternative forms of resilience in Greece", *Forschungsjournal Soziale Bewegungen*, 29(1): 50–61 (in German).

Koutrolikou, P., Spanou, D. (2013) "The local as an arena for emerging mobilizations and solidarity in the context of the current crisis", *Geographies*, 22: 52–66 (in Greek).

Krugman, P. (1991) *Geography and Trade*, Cambridge, MA: MIT Press.

Krugman, P. (1993) "Lessons of Massachusetts for EMU", in: Torres, F. and Giavazzi, F. (eds) *Adjustment and Growth in the European Monetary Union*, London: CEPR, Cambridge University Press.

Krugman, P. (2010) "The Euro Trap", *The New York Times*, 9 May 2010.

L'Ordine Nuovo, 1919–1920, (1995) Turin: Einaudi.

Labrianidis, L. (ed) (2008) *The Moving Frontier. The Changing Geography of Production in Labour-Intensive Industries*, Aldershot: Ashgate.

Labrianidis L. (2011) *Investing in leaving: The Greek case of international migration of professionals in the globalization era*. Athens: Kritiki (in Greek).

Labrianidis, L., Vogiatsis, N. (2013) "The mutually reinforcing relations between international migration of highly educated labour force and economic crisis: the case of Greece", *Southeast European and Black Sea Studies,* 13(4): 525–551.

Lafazani, O. (2004) "Grassroots migrant organizations and Antiracist Initiatives in Greece: difficult encounters, interesting connections". Paper presented in the International Symposium "*Transnational Europe I – Migration across southern/eastern borders*", University of Crete, Rethymno.

Lafazani, O. (2012) "The Border between Theory and Activism", *An International E-Journal for Critical Geographies*, 11(2): 189–193.

Lafazani, O. (2014) *Geographies of transnational migration*, PhD Thesis, Department of Geography, Harokopio University, Athens (in Greek).

Lanziani, A. (ed) (2003) *Metamorfosi urbane. I luoghi dell'immigrazione*, Pescara: Sala Editori.

Lapavitsas, C. (2013) *Profiting without Producing. How Finance Exploits Us All*, London: Verso.

Lapavitsas, C., Kaltenbrunner, A., Labrinidis, G., Lindo, D., Meadway, J., Michell, J., Painceira, J.P., Pires, E., Powell, J., Stenfors, A., Teles, N., Vatikiotis, L. (2012) *Crisis in the Eurozone*, London: Verso.

Lazzarato, M. (2012) *The Making of the Indebted Man. An Essay on the Neoliberal Condition*, Los Angeles: Semiotext(e).

Lefebvre, H. (1973) *La Survive du Capitalisme: la reproduction des rapports de production,* Paris: Antropos (translated in Greek 1975 and in English 1976).

Lefebvre, H. (1974) *La production de l' espace*, Paris: Anthropos.

Lefebvre, H. (1976) *De l' Etat* (4 volumes), Paris : Union générale d'éditions

Leonardi, R. (2006) "Cohesion in the European Union", *Regional Studies,* 40(2): 155–166.

Leontidou, L. (2012) "Athens in the Mediterranean 'movements of the piazzas'. Spontaneity in material and virtual public spaces", *City*, 16(3): 299–312.

Levi, C. (1990) *Cristo se è fermato a Eboli*, Torino: Einaudi.

LIVEWHAT (2016) *Living with Hard Times. How Citizens React to Economic Crises and Their Social and Political Consequences*, EU 7th Framework Programme (613237) in: www.livewhat.unige.ch/.

López Hernández, I., Rodríguez López, E. (2010) *Fin de ciclo: financiarización, territorio y sociedad de propietarios en la onda larga del capitalismo hispano (1959–2010)*, Madrid: Traficantes de Sueños.

Loureiro de Matos, F., Miramontes Carballada, A., Sa Marques, T., Ribeiro, D. (2015) "Housing in a time of crisis: Portugal and Spain an overview", www.enhr2015.com/images/Southern_European.

Lovering, J. (1999) "Theory led by policy: the inadequacies of the new regionalism", *International Journal of Urban and Regional Research,* 23(2): 379–395.

MacKinnon, D., Cumbers, A., Pike A., Birch K., MacMaster R. (2009) "Evolution in economic geography: institutions, political economy and adaptation, *Economic Geography*, 85(2): 804–829.

Magnifico, G. (1973) *European Monetary Unification,* London: Macmillan.

Mantanika, R., Kouki, H. (2011) "The spatiality of a social struggle in Greece at the time of the IMF: reflections on the 2011 mass migrant hunger strike in Athens", *City,* 15(3–4): 482–490.

Marques, P. (2015) "Why did the Portuguese economy stop converging with the OECD? Institutions, politics and innovation", *Journal of Economic Geography*, 15(5): 1009–1031.

Marsden, T., Banks, J., Bristow, G. (2000) "Food supply chains approaches: exploring their role in rural development", *Sociologia Ruralis,* 40(4): 424–438.

Mart, M. (2013) "Housing bubble, crisis and social struggle in Spain", *Geographies*, 22: 45–48.

Martin, R. (2000) "EMU versus the Regions? Regional convergence and divergence in Euroland", Working Paper No. 179, ESRL Center for Business Research, University of Cambridge.

Martinez-Veiga, U. (2001) *El Ejido. Discriminación, exclusión y racismo*, Madrid: Los Libros de la Catarata.

Massey, D. (1999) *Power-geometries and the Politics of Space-Time,* Hettner-Lectures 2, Heidelberg: University of Heidelberg.

Massey, D. (2004) "Geographies of responsibility", *Geografiska Annaler*, 86 B(1): 5–18.

Massey, D. (2006) "Space, Time and Political Responsibility in the Midst of Global Inequality", *Erdkunde – Archive for Scientific Geography* (Department of Geography, University of Bonn, Germany) 60(2): 93–106.

Massey, D. (2007) *World City*, London: Polity.

Massey, D. (2012) *Radical Spatiality and the Question of Democracy*, Invited Talk, Ceremony for Doreen Massey's Honorary Doctorate at the Geography Department, Harokopio University, Athens, Greece, Athens: Harokopio University.

Massey, D. (2015) "Vocabularies of the economy", in Hall, St., Massey, D., Rustin, M. (eds) *After Neoliberalism? The Kilburn Manifesto*, London: Soundings, pp. 24–36.

Massey, D., Hall, S. (2010) "Interpreting the crisis", *Soundings,* www.fags.org/periodicals/201004.

Mayayo, G. (2007) "The Spanish mortgage market and the American subprime crisis" www.ahe.es/bocms/images (accessed September 2013).

Mayer, M. (2013) "Against and beyond the crisis: the role of urban social movements", *Geographies*, 22: 67–72.

McLeod, G. (2001) "New regionalism reconsidered: globalization and the remaking of political economic space", *International Journal of Urban and Regional Research,* 25(4): 804–829.

Medelfart, K.-H., Overman, H., Venables, A. (2003) "Monetary union and the economic geography of Europe", *Journal of Common Market Studies*, 41(5): 847–868.

Merrifield, A., Swyngedouw, E. (eds) (1996) *The Urbanization of Injustice*, London: Lawrence & Wishart.

Mezzadra, S., Neilson, B. (2013) *The Border as a Method or the Multiplication of Labor*, Durham and London: Duke University Press.

Miguélez, F., Recio, A. (2010) "The uncertain path from the Mediterranean welfare model in Spain", in: Anxo, D., Bosh, G., Rubery, J. (eds) *The Welfare State and Life Transitions. A European Perspective*, Cheltenham: Edward Elgar, pp. 284–308.

Mingione, E. (1991) *Fragmented Societies. A Sociology of Economic Life beyond the Market Paradigm*, Oxford: Blackwell.

Mingione, E. (1995) "Labour market segmentation and informal work in Southern Europe", *European Urban and Regional Studies*, 2(2): 121–143.

Mingione, E. (1998a) "The social and historical construction of the models of industrial development", in *Space, Inequality and Difference. From "Radical" to "Cultural" Formulations*? Milos, 1996, Seminars of the Aegean, Atene, 1998, pp. 82–106.

Mingione, E. (1998b) *Sociologia della vita economica,* Roma: Carocci.

Mingione, E. (2000) "Modello Sud Europeo di welfare. Forme di povertà e politiche contro l'esclusione sociale" *in Sociologia e Politiche Sociali* (P. Donati ed) a.3, n.1, pp. 87–112.

Mingione, E. (2009) "Family, welfare and districts: the local impact of the new migrants in Italy", *European Urban and Regional Studies*, 16(3): 225–230.

Morgan, K. (1977) "The learning region: institutions, innovation and regional renewal", *Regional Studies,* 31(5): 491–503.

Morrissey, J. (2009) "Lessons in American geopolitik: Kaplan and the return of spatial absolutism", *Human Geography*, 2(2): 37–39.

Nardone, C. (1971) *Il pensiero di Gramsci*, Bari: De Donato.

Nesi, E. (2010) *Storia della mia gente*, Milano: Bombiani.

Nicevero, A. (1898) *L' Italia barbara contemporanea*, Milano/Palermo.

Nurra, M., Azzu, M. (2011) *Asinara Revolution*, Milano: Bompiani.

Observatorio Metropolitano. (2011a) *Crisis y revolución en Europa. People of Europe, rise up!* Madrid: Traficantes de Sueños. www.rebelion.org/docs/138745.pdf. IN ENGLISH.

Observatorio Metropolitano. (2011b) *La crisis que viene. Algunas notas para afrontar esta década.* Madrid: Traficantes de Sueños. www.traficantes.net/sites/default/files/pdfs/La%20crisis%20que%20viene-Traficantes%20de%20Sueños.pdf.

Observatorio Metropolitano. (ed) (2013) *Paisajes devastados. Después del ciclo inmobiliario: impactos regionales y urbanos de la crisis*, Madrid: Traficantes de Sueños.

OECD (2003) Annual Report, https://oecd.org/about/2506789.

OECD. (2010) *Labour Productivity*, https://data.oecd.org/lprdty/gdp-per-hour-worked.htm.

OECD. (2013) *International Migration Outlook*, Paris: OECD.

Paci, M. (1972) *Mercato del lavoro e classi sociali in Italia*, Bologna: Il Mulino.

Painter, J. (2003) "Towards a post-disciplinary political geography", *Political Geography*, 22: 637–639.

Papadopoulou, E., Sakellaridis, G. (eds) (2012) *The Political Economy of Public Debt and Austerity in the EU*, Brussels: Transform!.

Papataxiarchis, E. (2016a) "Being 'there': at the front line of the 'European refugee crisis'", Part 1, *Anthropology Today*, 32(2): 5–9.

Papataxiarchis, E. (2016b) "Being 'there': at the front line of the 'European refugee crisis'", Part 2, *Anthropology Today*, 32(3): 3–7.

Papataxiarchis, E. (2016c) "Unwrapping solidarity? Society reborn in austerity", *Social Anthropology*, 24(2): 205–210.

Papic, M., Reinfrank, R., Zeihan, P. (2010) "Insights economics: Greece and Germany to exit the Eurozone", *Straftor Documents,* www.marketoracle.co.uk/.

Peck, J. (2012) "Austerity urbanism", *City*, 16(6): 626–655, doi:10.1080/13604813.2012.734071.

Peck, J., Theodore, N., Brenner, N. (2010) "Postneoliberalism and its malcontents", *Antipode,* 41(supplement 1): 94–116.

Peet, R. (1985) "The social origins of environmental determinism", *Annals of the Association of American Geographers*, 75(3): 309–333.

Perrons, D. (2004) *Globalization and Social Change,* London: Routledge.

Perrons, D. (2010) *Why socio-economic inequalities increase? Facts and policy responses in Europe*, Directorate-General for Research 2010 Socio-economic Sciences and Humanities EUR 24471 EN.

Perrons, D. (2012a) "'Global' financial crisis, earnings inequalities and gender: towards a more sustainable model of development", *Comparative Sociology,* 11: 202–226.

Perrons, D. (2012b) "Regional performance and inequality: linking economic and social development through a capabilities approach", *Cambridge Journal of Regions*, 5: 15–29.

Petmesidou, M., Pavolini, E., Guillén, A.M. (2015) "South European healthcare systems under harsh austerity: a progress-regression mix?", in: Petmesidou, M., Guillén, A.M. (eds) *Economic Crisis and Austerity in Southern Europe*, London: Routledge, pp. 37–58.

Pickles, J., Smith, A. (2011) "Delocalization and the persistence in the European clothing industry", *Regional Studies*, 45: 167–185.

Pickles, J., Smith, A. (2015) *Articulations of Capital: Global Production Networks and Regional Transformations*, Chichester: Wiley.
Pike, A. (2007) "Whither Regional Studies?" Editorial, *Regional Studies,* 41(9): 1143–1148.
Pike, A., Rodriguez-Pose, A., Tomaney, J. (2007) "What kind of local and regional development and for whom?" *Regional Studies,* 41(9): 1253–1269.
Pivetti, M. (1998) "Monetary versus political unification in Europe", *Review of Political Economy*, 10(1): 5–25.
Plank, L., Staritz, C. (2015) "Global competition, institutional context and regional production networks: up-and downgrading experiences in Romania's apparel industry", *Cambridge Journal of Regions, Economy and Society*, doi:10.1093/cjires/rsv014.
Poço, M., Lopez, C., Silva, A. (2015) "Perceptions of tax evasion and tax fraud in Portugal", www.gestaodefraude.eu.
Politi, J. (2016) *Demonizing and Idealizing the South. A Literary Voyage in Space and Time*, Nicos Poulantzas Institute, Athens: Nisos Publ. (in Greek).
Portaliou, E. (2007) "Anti-global movements reclaim the city", *City,* 11(2): 165–175.
Portaliou, E. (2008) "Urban movements in Athens. Notes for postgraduate studies, NTUA", www.courses.arch.ntua.gr/121473.html (accessed 30 February 2013).
Porter, M. (1998) "Clusters and the new economics of competition", *Harvard Business Review*, 76(6): 77–90.
Poulantzas, N., Althusser, L., Balibar, E. (1980) *Debating the State*, Athens: Agonas (in Greek).
Pugliese, E. (1985) "Farm workers in Italy: agricultural working class, landless peasants or clients of the welfare state?", in: Hudson, R., Lewis, J. (eds) *Uneven Development in Southern Europe*, London: Methuen, pp. 123–139.
Pugliese, E. (1995) "New international migrations and the "European Fortress", in: Hadjimichalis, C., Sadler, D. (eds) *Europe at the Margins. New Mosaics of Inequality*, Chichester: J. Wiley, pp. 51–68.
Pugliese, E. (2002) *L' Italia tra migrazioni internationali e migrazioni interne*, Bologna: il Mulino.
Putnam, R. (1993) *Making Democracy Work,* Princeton: Princeton University Press.
Rakopoulos, Th. (2015) "Solidarity's tensions: informality, sociality and the Greek crisis", *Social Analysis*, 59(3): 85–104.
Recio, A. (2010) "Capitalismo español: la inevitable crisis de un modelo insostenibile", *Revista Economia Crítica*, 9: 198–222.
Reyneri, E. (2001) *Migrants in irregular employment in the Mediterranean countries of the European Union,* International Migration Papers, no 41, ILO, Geneva. www.Ilo.org/public/english/migrant/research.
Rodrigues, J., Reis, J. (2012) "The asymmetries of European integration and the crisis of capitalism in Portugal", *Competition and Change*, 16: 188–205.
Romão, T.G. (2015) *Evolution of the Portuguese construction sector*, Master Thesis, Tecnico Lisboa.
Rossi, U. (2009) "Growth Poles-Growth Centers", *International Encyclopedia of Human Geography*, Regional Development Section, Elsevier. www.elsevier.com/HUGY.
Roth, K.H. (2013) *Before the Greek Debt Crisis, a Pamphlet*, London: Zero Books.
Rozakou, K. (2015) "The Lesvos pass: crisis, humanitarian governance and solidarity", *Sychrona Themata*, 130–131: 13–16 (in Greek).

Rozakou, K. (2016) "Crafting the volunteer: voluntary association and the reformation of solidarity", *Journal of Modern Greek Studies*, 34: 78–101.

Said, E. (1979/1996 Greek edition) *Orientalism*, Athens: Nefeli (in Greek).

Said, E. (2001) "Orientalism reconsidered", in: *Reflection on Exile and other Literary and Cultural Essays*, New York: Granta books (Greek translation, Athens: Scripta, 2006), pp. 317–342.

Salento, A. (2014) "The neo-liberal experiment in Italy: false promises and social disappointments", CRESC Working Paper Series, Working Paper No. 137, www.cresc.ac.uk.

Sayer, A. (2015) *Why We Can't Afford the Rich*, Bristol: Polity Press.

Schmidt, I. (2010) "European capitalism: varieties of crisis", *Alternate Routes – A Journal of Critical Social Research*, 22: 71–86.

Schraad-Tischler, D., Kroll, C. (2014) and (2016) *Social Justice in the EU-A cross-national comparison*, Bertelsmann Stiftung, Bielefeld. www.bertelsamnn-stiftung.de.

Schultz-Dornburg, J. (2012). *Ruinas modernas. Una topografía del lucro.* Barcelona: Àmbit.

Scott, A., Storper, M. (1988) "The geographical foundations and social regulation of flexible production complexes", in: Wolch, J., Dear, M. (eds) *The Power of Geography,* London: Allen and Unwin, pp. 21–40.

Scroccaro, A., Sivieri, C. (2009) "Timişoara e l'imprenditoria della calzatura veneta. Dal distretto dello SportSystem di Montebelluna a 'Trevişoara', in: Bertoncin, M., Marini, D., Pase, A. (eds) *Frontiere Mobili. Delocalizzazione e internationalizzazione dei territori produttivi veneti,* Venezia: Marcilio, pp. 71–94.

Sen, A. (2009) *The Idea of Justice*, Cambridge, MA: Harvard University Press.

Sevilla-Buitrago, A. (2015a) "Espacialidades indignadas: la producción del espacio público en la #spanishrevolution", *ACME: An International Journal for Critical Geographies*, 14(1) (online open access).

Sevilla-Buitrago, A. (2015b) "Crisis and the city. Neoliberalism, austerity planning and the production of space", in: Echardt, F., Ruiz-Sánchez, J. (eds) *City of Crisis. The Multiple Contestation of Southern European Cities*, Bielefeld: Transcript Verlag, pp. 31–49.

Siatitsa, D. (2014) *Claims for the right to housing in cities of Southern Europe: the discourse and the role of social movements*, PhD Dissertation, Department of Urban Planning, NTUA, Athens (in Greek).

Silva, R. (2013) "Housing crisis in Portugal", *Geographies*, 22: 49–51.

Simonazzi, A., Ginzburg, A., Nocella, G. (2013) "Economic relations between Germany and southern Europe", *Cambridge Journal of Economics*, 37: 653–675.

Skidelsky, R. (2013) "Four Fallacies of the second Great Depression", www.project-syndicate.org/commentary/robert-skidelsky-explains.

Skordili, S. (1999) *Geographical restructuring of manufacturing: the case of the food sector in Greece*, PhD Dissertation, Department of Planning and Regional Development, AUTH (in Greek).

Skordili, S. (2013) "Economic crisis as a catalyst for food planning in Athens", *International Planning Studies*, 18(1): 129–141.

Smith, A. (2013) "Europe and an inter-dependent world: uneven geo-economic and geo-political developments", *European Urban and Regional Studies*, 20(1): 3–13.

Smith, A. (2015) "Macro-regional integration, the frontiers of capital and the externalization of economic governance", *Transactions of the IBG*, 40(4): 507–522.

Smith, A., Rainnie, A., Dunford, M., Hardy, J., Hudson, R., Sadler, D. (2002) "Networks of value, commodities and regions: reworking divisions of labour in macro-regional economies", *Progress in Human Geography,* 26(1): 41–63.

Smith, G. (1999) *Confronting the Present: Towards a Politically Engaged Anthropology,* Oxford: Berg.

Soja, E. (2010) *Seeking Spatial Justice,* Minneapolis: University of Minnesota Press.

Soltas, E. (2015) "The Madre of All Bubbles", http://evansoltas.com/2015/04/03/the-madre-of-all-bubbles/.

Sotiropoulou, I. (2012) *Exchange networks and parallel currencies: theoretical approaches and the case of Greece,* PhD Dissertation, Department of Economics, University of Greece: Rethymno (in Greek).

Spourdalakis, M. (1988) *The Rise of the Greek Socialist Party,* London: Routledge.

Standing, G. (2016) *The Corruption of Capitalism. Why Rentiers Thrive and Work Does Not Pay,* London: Bite back Publication.

Stavridis, St. (2017) "Life after the squares: reflections on the consequences of the occupy movements" (intervention on Syntagma), *Social Movements Studies,* 16(1): 119–151.

Stockhammer, E. (2014) "The Euro crisis and contradictions of neoliberalism in Europe", *Post Keynesian Economics Study Group,* Working Paper 1401, www.postkeynesian.net.

Storper, M. (1997) *The Regional World: Territorial Development in a Global Economy,* New York: Guilford.

Sweezy, P. (1994) "The triumph of financial capital", *Monthly Review,* 46(2): 1–11.

Swyngedouw, E. (1977a) "Excluding the other: the production of scale and the scaled politics", in: Lee, R., Wills, J. (eds) *Geographies of Economies,* London: E. Arnold, pp. 167–176.

Swyngedouw, E. (1977b) "Neither global nor local: "glocalization" and the politics of scale", in: Cox, K. (ed) *Spaces of Globalization,* New York: Guilford, pp. 137–166.

Swyngedouw, E. (2000) "Authoritarian governance, power and the politics of rescaling", *Environment and Planning D: Society and Space,* 18: 63–76.

Taibo, C. (2013) "The Spanish indignados: a movement with two souls", *European Urban and Regional Studies,* 20(1): 155–158.

Tarpagos, A. (2010) "The construction sector: from the 'golden decade' to over-accumulation crisis and disaster in Greece", *Theseis,* 113: 33–41 (in Greek).

Thirwall, T. (2000) *The Euro and Regional Divergence in Europe,* London: New Europe Research Trust.

Thoidou, E., Foutakis, D. (2006) "Metropolitan Thessaloniki and urban competitiveness: programming, transformation and implementation of a 'vision' for the city", *Geographies,* 12: 25–46 (in Greek).

Todl, G. (2000) "EU regional policy in southern periphery: lessons for the future", *South European Society and Politics,* 3(1): 93–129.

Trikliminiotis, N., Parsanoglou, D., Tsianos, V. (eds) (2015) *Mobile Commons, Migrant Digitalities and the Right to the City,* London: Palgrave MacMillan.

Vaiou, D. (1997) "Informal cities? Women's work and informal activities on the margins of the European Union", in: Lee, R., Wills, J. (eds) *Geographies of Economies,* London: Arnold, pp. 321–330.

Vaiou, D. (2014) "Tracing aspect of the Greek crisis in cities: putting women in the picture", *European Urban and Regional Studies,* 23(3): 220–230.

Vaiou, D., Hadjimichalis, C. (2004) *With the sewing machine in the kitchen and the Poles in the fields. Cities, regions and informal work*, Athens: Exandas (2nd editions, in Greek)

Vaiou, D., Hadjimichalis, C. (2012) *Space in Left Thought*, Athens: Nissos/N. Poulantzas Institute (in Greek).

Vaiou, D., Kalandides, A. (2016) "Practices of collective action and solidarity: reconfigurations of the public space in crisis-ridden Athens, Greece", *Journal of Housing and the Built Environment*, 31: 457–470.

Vaiou, D., Kalandides, A. (2017) "Practices of solidarity in Athens: reconfigurations of public space and urban citizenship", *Citizenship Studies*, 21(4): 440–454.

Vaiou, D. et al. (2007) *Intersecting Patterns of Everyday Life and Socio-Spatial Transformations in the City. Migrant and Local Women in the Neighbourhoods of Athens*, Athens: L-Press and NTUA.

Valverde, C. (2013) *No nos lo creemos. Una lectura crítica del lenguaje neoliberal*, Barcelona: Icaria.

Van Vossole, J. (2014) "Framing PIGS to clean their own stable". Paper presented at 7th ECPR Conference, Bordeaux, France.

Varoufakis, Y. (2011) *The Global Minotaur. America, Europe and the Future of the Global Economy*, London: Zed Books.

Varoufakis, Y. (2016) *And the Weak Suffer What They Must? Europe's Crisis and America's Economic Future*, New York: Nation Books.

Vatavali, F., Chatzikonstantinou, E. (2015) *Geographies of energy poverty in Athens in the context of the crisis.* (in Greek) www.latsis-foundation.org/eng/education-science-culture/science/grants/scientific-projects/2015/geographies-of-energy-poverty-in-athens-in-the-context-of-crisis.

Vatavali, F., Koutrolikou, P., Balabanidis, D., Siatitsa, D. (2013) "Crisis regimes and emerging social movements in Southern Europe: urban development, housing and local struggles", *Geographies*, 22: 33–34 (in Greek).

Vathakou, E. (2015) "Citizens' solidarity initiatives in Greece during the financial crisis", in: Huliaras, A., Clark, J. (eds) *Austerity and the Third Sector in Greece: Civil Society at the European Front Line*, Aldershot: Ashgate, pp. 167–192.

Vàzquez-Barquero, A. (1992) "Local development and flexible accumulation: learning from history and policy", in: Garofoli, G. (ed) *Endogenous Development and Southern Europe*, Aldershot: Avebury, pp. 31–48.

Velegrakis, G. (2015) "Cold mining in Chalkidiki and local struggles: analysis of voting practices in municipal and national elections", *Geographies*, 26: 77–92 (in Greek).

Venables, A. (1996) "Equilibrium locations of vertically linked industries", *International Economic Review*, 37: 341–359.

Venturini, A. (1988) "An interpretation of Mediterranean migrations", *Labour*, 2: 125–154.

Verducci, F. (2003) "Immigrazione e mutamento sociale. Lavoro ed integrazione sociale nel distretto industriale della calzatura", Universita deglli Studi di Macerata, Dip. Di Sociologia (mimeo).

Vradis, A., Dalakoglou, D. (2011) *Revolt and Crisis in Greece,* Oakland/Edinburgh: AK Press.

Wallerstein, I. (1983) *Historical Capitalism*, London: Verso.

Weeks, J. (2014) "Euro crises and Euro scams: trade not debt and deficits tell the tale", *Review of Political Economy*, 26(2): 171–189.

West, R. (1982) *Black Lamb and Grey Flacon: A Journey through Yugoslavia*, New York: Penguin.

Williams, A. (ed) (1984) *Southern Europe Transformed*, Cambridge: Harper&Row.

Wilson, J., Swyngedouw, E. (2015) "Seeds of dystopia: post-politics and the return of the political", in: Wilson, J., Swyngedouw, E. (eds) *The Post-Political and Its Discontents,* Edinburgh: Edinburgh University Press, pp. 1–22.

Ybarra, J.A., San Miguel, B., Hurtado, J., Santa Maria, J. (2004) *El calzado en el Vinalopó entre la condinuidad y ruptura*, University Alicante: Alicante.

Zavos, A., Koutrolikou, P., Siatitsa, D. (2017) "Changing landscapes of urban citizenship: Southern Europe in times of crisis", *Citizenship Studies*, 21(4): 379–392.

Index